AF556460

PHOTOSYNTHESIS
IN
PLANTS

PHOTOSYNTHESIS IN PLANTS

By

Dr. Shubhrata R. Mishra
Department of Botany
Vikram University
Ujjain (M.P)

DISCOVERY PUBLISHING HOUSE
NEW DELHI-110002

First Published-2004

ISBN 81-7141-891-0

Published by

DISCOVERY PUBLISHING HOUSE
4831/24, Ansari Road, Prahlad Street,
Darya Ganj, New Delhi-110002 (India)
Phone: 23279245 • Fax: 91-11-23253475
E-mail:dphtemp@indiatimes.com

Printed at :
ARORA OFFSET PRESS
Laxmi Nagar, Delhi-92.

PREFACE

The present title "Photosynthesis in Plants" is a classical branch in plant physiology. Biochemists purify photosynthetic enzymes and study their characteristics in the test tube; biophysicists isolate photosynthetic membranes and determine their spectroscopic properties in cuvettes; molecular biologists clone the genes that encode photosynthetic proteins and study their regulation during development. In contrast, plant physiologists study photosynthesis in action at different levels of organization, including the chloroplast, the cell, the leaf and the whole plant. Stated differently, biochemists, biophysicists and molecular biologists study cellular components more or less in isolation, whereas plant physiologists investigate the way in which the components interact with each other to carry out biological processes and functions.

The need to understand how the components of living organisms interact with each other will not be eliminated by a detailed characterization of the genome. On the contrary, the enriched understanding of gene regulation in plants will generate a wealth of new questions concerning plant function. It is hoped that the undergraduates of today will receive from this textbook both the knowledge and the motivation to become the scientists who will lead plant physiology into the 21st century.

The author expresses his thanks to his friends and colleagues whose continuous inspirations have invited him to bring out this text.

The author expresses his gratitute to Mr. Wasan and staff of M/s Discovery Publishing House for their whole hearted co-operation in the publication of this book.

Constructive criticisms and suggestions for improvement of the book will be thankfully acknowledged.

Author

CONTENTS

1

Photophysiology

The radiation from the sun produces a direct heating effect, and also produces photochemical transformations. After these transformations have been completed, the energy appears in the form of heat. Accordingly, all the sun's energy eventually ends up as beat, but certain portions of the radiation from the sun take part in vital photochemical processes before becoming heat. Light as an ecological factor is generally highly directional. It differs in this respect from the temperature factor since heat often reaches the living organism from many different directions at the same time. Light is extremely variable. It changes over a tremendous range, often very rapidly.

Many organisms can respond to a value of light that is only one ten billionth of full sunlight. Although we often think of temperature as varying widely on the earth's surface, the magnitude of its variation is extremely small compared to that of light. Light can change from a value near zero to its maximum within a few hours. Light is essential for most plants and animals, though some can do without it. For the continued existence of organisms two requirements must be met. First, light must not be so strong as to cause serious harm at any stage in the life history. We shall see that a considerable variation exists in the upper, limits of tolerance to the light factor. Second, for those animals and plants that require light it must be sufficient in intensity and in duration. The intensity of the light must be above the thresh old for the organism concerned and the total amount of light received during the period when it is needed must be adequate.

Distribution of Light

In discussing the distribution of light it is convenient to start with sunlight as the chief source and trace the radiation as it passes through the air and water environments. The magnitude of the solar radiation as it reaches the outer atmosphere referred to as the *solar constant* has the value of about 1.9 g-cal/cm^2/min. At sea level the intensity of solar radiation averages about 1.5 g-cal/cm^2/min. If the radiation received from the sun were evenly distributed over the surface of the globe it would be sufficient to melt a layer of ice 35 m thick during the course of a year. At a latitude of 44°N the energy received on the earth's surface from the sun is equivalent to the light that would be produced by hanging a 250-watt lamp over each square meter of the ground. We know very well however that the radiation from the sun is not evenly distributed either in time or in space.

We wish to inquire what its variations both in quality and in quantity and in Light changes in spectral distribution and in angular distribution. The light factor also varies in intensity and in duration resulting in differences in the total amount of light falling on each unit of surface for each unit of time, such as a day or a month. The changes in the light factor in these respects will he examined as they affect the terrestrial and the aquatic environments.

Light on Land

Spectral composition

The spectral distribution of light as it reaches the earth's surface differ as to the exact wavelength limits to be assigned to the different portions of the spectrum. Roughly speaking radiation of wavelength longer than 7600 Angstrom units is considered to be infrared, and that of wavelength shorter than about 3600Å is designated as ultraviolet. Almost one-half of the total emission of the sun is infrared radiation, and almost one-half is visible light. These proportions remain approximately the same at the earth's surface, regardless of the total intensity of sunlight. The ultraviolet component, however, is always only a small percentage of the total radiation, and it may be reduced to immeasurably small quantities under certain' circumstances.

Intensity of light

The intensity of light reaching the earth's surface varies with the angle of incidence and with the amount of absorption by the atmosphere and by obscuring features. The lower the altitude of the sun, the smaller is the angle of incidence and the longer is the path of the light through the atmosphere with corresponding reduction in intensity. Changes in the sun's altitude result from differences in latitude as well as from changes in the season and in the time of day. The greatest intensity of sunlight occurs at positions on the earth's surface and at times at which the sun is most nearly overhead. At higher latitudes, the intensity of light is correspondingly reduced. At 50°N latitude, for example, during the period of the equinox in March and in September when the day is everywhere 12 hours long, the intensity of sunlight is only about one half of what it is at the equator. Latitude thus has a definite effect; but other factors may have much greater influence upon the light factor.

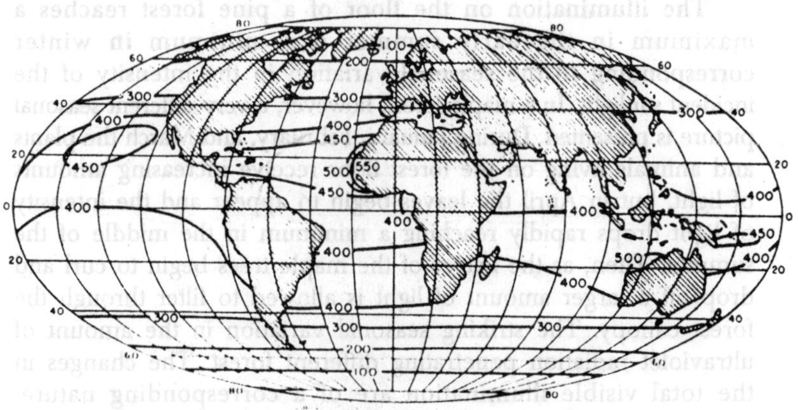

Fig. 1.1. Total solar radiation (g-cal/day/cm² of horizontal surface) on March 21 with average cloudiness.

Moisture, clouds and dust in the atmosphere have a profound and irregular effect in reducing illumination Living organisms may also act to diminish the intensity of daylight, as is clearly shown by the forest vegetation, and this represents a reciprocal action in which the inhabitant; modify the light factor in their own environment. Different forest communities vary widely in the degree to which they diminish the sun's radiation. Cottonwood (popular) trees tend to grow rather widely spaced, and the relatively sparse foliage allows many patches of sunlight to reach the ground. The canopies of pines and oaks usually have fewer gaps. Measurements made in Illinois revealed that the portion of the forest floor exposed to direct sunlight was 84 per cent for popular, 77 per cent for pine, and 35 per cent for oak. In elm-maple forests and in tropical rain forests the canopy typically is complete, with the result that none of the direct sunlight reaches the forest floor.

Other measurements showed that the light on the forest floor which bad passed through the leaves of the canopy was reduced in intensity to 3.5 per cent of value above the tree tops in the oak forest; corresponding figures for elm-maple and rain forests were 0.4 per cent and 0.2 per cent, respectively. Thus less than 1 per cent of the light outside the forest reaches the ground if the canopy is complete. Different types of forests exhibit different seasonal influences on the light factor. Underneath a stand of pine trees the light is reduced by about the same amount throughout the year since the trees are evergreen.

The illumination on the floor of a pine forest reaches a maximum in the early summer d a minimum in winter corresponding to the seasonal variation in the intensity of the incident sunlight. In a maple forest, however, a very different seasonal picture is presented. During January, February, and March the plants and animals living on the forest floor receive increasing amounts of light, but in April the leaves begin to appear and the intensity of light drops rapidly reaching a minimum in the middle of the summer. Then, as the leaves of the maple trees begin to curl and drop off a larger amount of light is allowed to filter through the forest canopy. The striking seasonal variation in the amount of ultraviolet radiation penetrating different forest. The changes in the total visible illumination are of a corresponding nature. Accordingly, if you were a beetle, or, if you prefer, a lily growing in a maple forest, the brightest month of the year for you would be April and the darkest would be July. In this situation living organisms have modified the environment in respect to the light factor so that the seasons have been practically reversed.

Duration and amount of light

The total amount of light received by an organism is determined both by the intensity of be light and the duration of the period of irradiation. The situation is similar to the exposure of a photographic plate the amount of blackening of the plate is determined by the intensity of the light times the period of exposure. In natural habitats, the variation in the length of day often has a greater effect on the total amount of light received

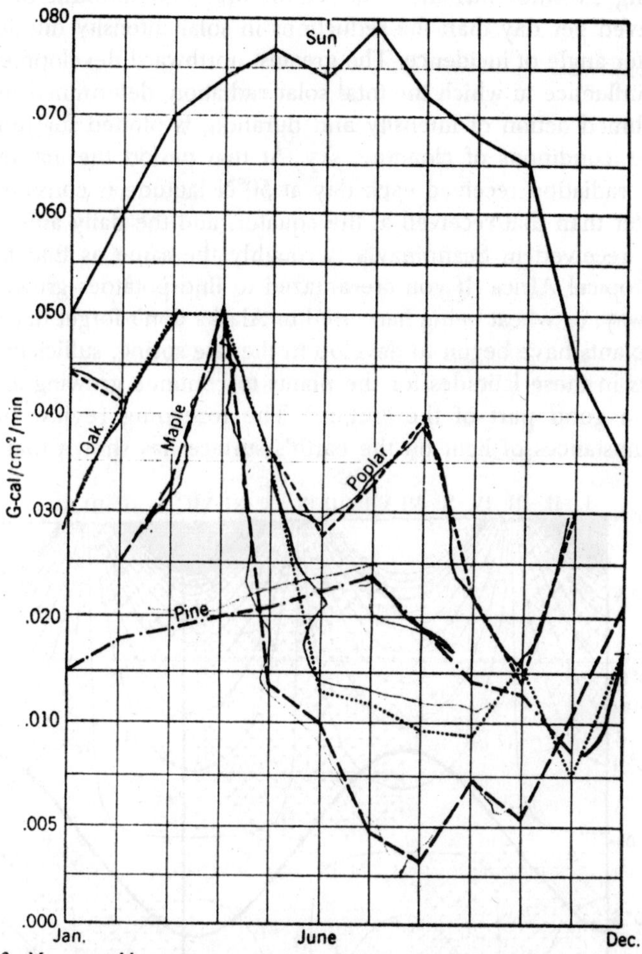

Fig. 1.2. Mean monthly maximum intensities of solar ultraviolet radiation (2900–4000A) in forest communities of northern Indiana.

than do differences in the intensity of the sun at noon. On the equator the day is always 12 hours long, but in the temperate regions the day grows longer as spring progresses. This effect is accelerated at higher latitudes, and the day becomes 24 hours long during the summer in the polar regions. The day becomes correspondingly shorter after the summer solstice.

Up to moderately high latitudes the increase in length of day during summer has more effect on the total amount of light received per day than the reductions in solar intensity due to the greater angle of incidence. The greatest northward development of this influence in which the total solar radiation, determined by the combined action of intensity and duration, is plotted for June 21 under conditions of cloudless sky. At that period the amount of solar radiation received each day at 50°N latitude is considerably greater than that received at the equator, and the daily amount of light received in Scandinavia is roughly the same as that falling on tropical Africa. If you are amazed to find potatoes growing in Norway, or wheat being harvested in Alaska don't forget that once the plants have begun to develop in the late spring, sufficient light exists in these latitudes for the plants to continue growing all day and a good part of the "night." The foregoing review of the circumstances of light on the earth's surface has shown that light

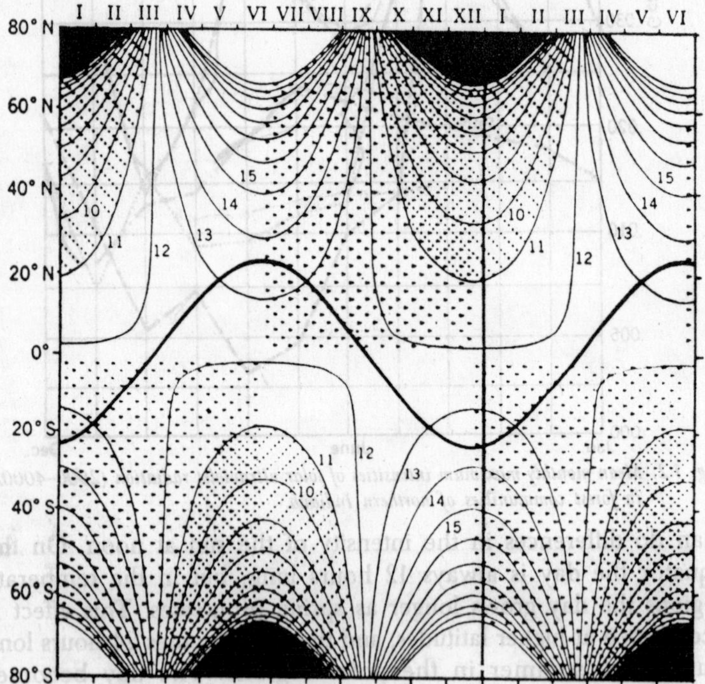

Fig. 1.3. Variation in length of day at the indicated latitudes during the indicated months. Arrowheads indicate periods of decreasing daylength. The heavy line shows the course of zenith sun.

is everywhere sufficient for life of some type at least for part of the year. Plants and animals exist from pole to pole. Mosses and lichens grow in abundance right up to the edge of the ice in the polar regions.

On the other hand, light is nowhere too strong for animals and plants of some sort. Thus the whole range of light on the earth's surface is generally compatible with life. Of course, light may sometimes be too strong or too weak for individual species, and in these instances it controls growth and distribution.

Wavelength and Energy

Light is often defined as radiation perceived by the human eye. Although such a definition may be strictly correct, the word *light* is frequently used to refer to a larger class of elector magnetic waves. In this section we will discuss the range of electromagnetic radiation that is important in plant and animal physiology including the subdivisions into various wavelength intervals. The wavelength of light will be related to its energy. After noting the units for radiant flux and illumination, we will briefly consider some of the characteristics of solar radiation reaching the earth.

Light waves

Our concern here is with the regular and repetitive changes in the intensity of the minute electric and magnetic fields which indicate the passage of light, wave. Light can travel in a solid (glass), in a liquid (water), in a gas (air), and even in a vacuum (between the sun and the earth's atmosphere). One way to characterize light is by its wavelength- the separation in space between successive points of the same phase, such as between two successive peaks of a wave train. A wavelength is the distance per *cycle* of the wave. One unit for wavelengths of light is the Angstrom (Å), which equals 10^{-8} cm or 10^{-10} m. Another is the millimicron. However, the presently preferred unit for wavelengths of light is the nanometer (nm)-like the millimicron, it is 10^{-9} m and it will be used in this text (1 nm= 10Å).

The wavelength regions of major interest in biology are the ultraviolet, the visible, and the infrared (Table 1.1). Wavelengths below about 400 nm are referred to as ultraviolet (UV)–meaning on the other side of, or beyond violet, in the sense of having a shorter wavelength. The visible region extends from approximately 400 nm to 740 nm and is subdivided into various bands such as blue, green or red (Table 1.1). These divisions are based on the subjective colour experienced by humans. The infrared (IR) region has wavelengths longer than those of the red end of the visible spectrum, up to approximately 40 µm.

Table 1.1. Definitions and characteristics of the various wavelength regions of light. The ranges of wavelengths leading to the sensation of a particular colour are somewhat arbitrary and vary with individuals. Both frequencies and energies in the table refer to the particular wavelength indicated in column 3 for each wavelength interval. Wavelength magnitudes are those in a vacuum.

Colour	*Approximate wavelength range (nm)*	*Representative wavelength (nm)*	*Frequency (cycles/sec (Hertz)*	*Energy (ev per photon*	*Energy (Kcal/ "mole" of photons)*
Ultraviolet	below 400	254	11.30×10^{14}	4.88	112.5
Violet	400-425	410	7.31×10^{14}	3.02	69.7
Blue	425-490	460	6.25×10^{14}	2.70	62.7
Green	490-560	520	5.77×10^{14}	2.39	55.0
Yellow	560-585	580	5.17×10^{14}	2.14	49.3
Orange	585-640	620	4.84×10^{14}	2.00	46.2
Red	640-740	680	4.41×10^{14}	1.82	42.1
Infrared	above 740	1400	2.14×10^{14}	0.88	20.4

Besides wavelength, we can also characterize a light wave by its frequency of oscillation, v, and by a velocity of propagation that has the magnitude *v*. These three quantities are related as follows:

$$\lambda \nu = v \qquad ...(1)$$

where λ is the wavelength. For all wavelengths the speed of light in a vacuum is a constant, generally designated as *c*, which experimentally equals 299,792 ± 1km/sec, or 3.00×10^8 m/sec. Light passing through a medium other than a vacuum has a speed less than *c*. For example, light having a wavelength of 589 nm in a vacuum is decreased in speed 0.03% by air, 25% by water, and 40% by dense flint glass. The wavelength undergoes a decrease in magnitude equal to the decrease in the speed of light in these various media while the unchanging property of a wave propagating through different media is its frequency. Note that *v* is the frequency of the oscillations of the local electric and magnetic fields. For light, $\lambda_{vacuum}\nu$ equals v_{vacuum} by Equation 1, where V_{vacuum} is the constant *c*. Therefore, if we know the wavelength in a vacuum, we can calculate the frequency.

In fact, the wavelengths given for light generally refer to values in a vacuum, as is the case for columns 2 and 3 in Table 1.1 (λs in air differ only slightly from the magnitudes listed). As a specific example, let us select a wavelength in the blue region of the spectrum, *e.g.*, 460 nm. By equation I, the corresponding frequency equals $c/\lambda_{vacuum,}$ or $(3.00 \times 10^8$ m/sec)/(460×10^{-9}ml cycle), which is 6.52×10^{14} cycles/sec for such blue light. Since *v* does not change from medium to medium, it is often desirable to describe light by its frequency.

Light in Water

The sunlight available for plants and animals in the aquatic environment has entered the water from the air and hence has first been subjected to all the changes imposed upon it by the conditions above the surface. Ten per cent or more of the light is lost by reflection at the surface or in the special conditions just beneath the surface. In addition, in passing downward, the light is further modified by the water medium in respect to intensity, spectral composition, angular distribution, and time distribution.

Extinction and modification of light

The light factor in the water environment is subject to a number of variable influences that modify it profoundly. In order to consider the situation first in its simplest terms, imagine a lake from which we have pumped out all the natural water and refilled with distilled water. Our lake then contains water of uniform and maximum transparency from top to bottom. Light is reduced in intensity both by absorption and by scattering, and the rate of reduction is measured as the extinction rate, although the term absorption rate is sometimes used loosely for toe combined effect. Pure water causes the extinction of light at different rates in different parts of the spectrum. Measurements of illumination made with a photometer placed in a watertight case and lowered into our imaginary lake would reveal the changes. At a depth of 70 m the blue component of sunlight has been reduced to about 70 per cent of its intensity at the surface.

At the same depth the yellow component of sunlight has been reduced to 6 per cent of its incident value. The orange and red components have been extinguished very much more rapidly. At a

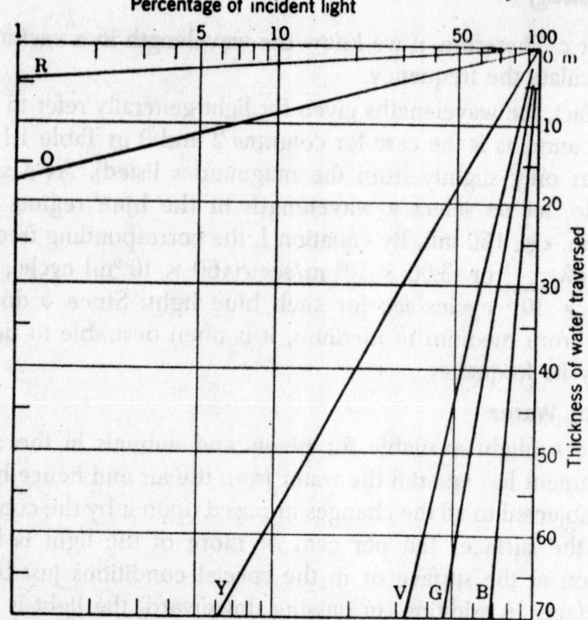

Fig. 1.4. Reduction in intensity (logarithmic scale) of the colour components of sunlight (indicated by initial letters) at increasing depths (linear scale) in a lake of optically pure water.

depth of 4 meters, red light has already been diminished to about 1 per cent of its surface intensity. Since we know that the daylight incident upon the surface of our lake differs markedly in intensity in different parts of the spectrum, we are dealing with the unequal extinction of an unequal spectrum. The extinction by pure water of light at the two ends of the spectrum is much more rapid than in the middle of the spectrum. The blue component of light is also the most penetrating in "the clearest ocean and lake waters. At depths of 100 m or more the blue light becomes completely predominating; this was observed directly by Beebe in his bathysphere dive off Bermuda.

An observer looking straight down into the water from the deck of a ship in the clear tropical parts of the ocean sees the blue colour resulting from this selective absorption. Since sunlight has been shorn progressively of its longer and shorter wavelengths, the only component remaining to be scattered upward again to the eye is the blue. Anyone who has seen the intense indigo of the tropical oceans will never forget it. The foregoing has shown that even pure water absorbs light at a very rapid rate compared to air

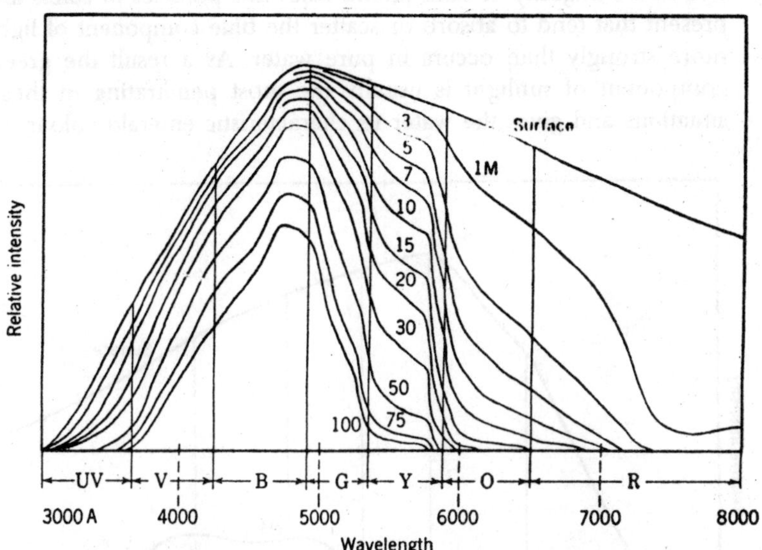

Fig. 1.5. The spectral distribution of solar energy at the earth's surface and after modification by passage through the indicated meters of pure water. Similar light conditions would be found in the clearest ocean and lake waters.

and causes a profound-change in spectral distribution. In the clearest parts of the ocean and in exceptionally clear lakes the optical properties of the water are closely similar to those of pure water.

Other natural waters contain suspended particles and dissolved material in sufficient quantities to cause a further reduction in transparency and a further alteration in spectral composition. Suspended material in the water includes living organisms that increase the extinction of the light and thus modify their own environment in this respect. Illumination is reduced by beds of kelp along the seacoast and by submerged or floating vascular plants along the shores of inland waters in much the same way that light is cut down by the larger vegetation on land.

In the free water of ponds, lakes, and the oceans the phytoplankton is sometimes sufficiently abundant to produce a noticeable reduction of light. Plankton populations may cause an additional extinction of light indirectly by adding detritus or stains to the water after the organisms have died and disintegrated. A thick "bloom" of algae in a pond may thus reduce the light supply to such an extent as to curtail its own growth and that of other plants in the water layers beneath. In temperate and coastal seas and in the majority of clear inland lakes fine particles or stains are present that tend to absorb or scatter the blue component of light more strongly than occurs in pure water. As a result the green component of sunlight is usually the most penetrating in these situations and gives the water its characteristic emerald colour.

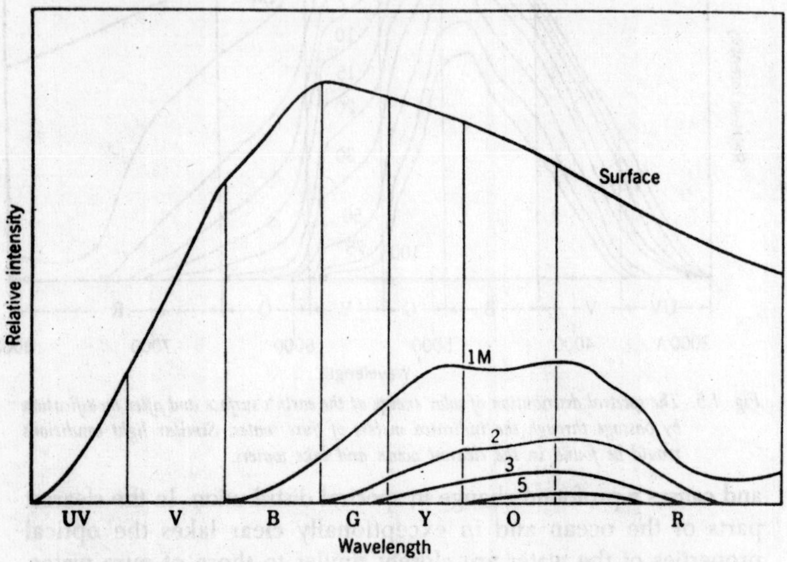

Fig. 1.6. Reduction in intensity and shift in spectral composition of light in heavily stained Rudolph Lake, Wisconsin.

The organic stains occurring in some ponds and rivers absorb the shorter wavelengths so strongly that the red or orange components of sunlight become the most penetrating. In Rudolph Lake, Wisconsin, for example, the combined absorption of the water and of stains in the water causes not only a very rapid reduction in the light with depth but also a shift in the-position of the maximum light to the red region of the spectrum. Since natural waters differ very greatly in transparency and in selective absorption, a study of the penetration of each part of the spectrum in each body of water would have to be made for a complete description of the light conditions.

An approximate comparison of the illumination in various natural waters can be made on the basis of the relative transparency to the central part of the spectrum. Sample values are given in Table 1.2 for representative bodies of water. Since the great differences in the extinction of light per meter have a cumulative effect, an even greater contrast is presented in the depths at which a given fraction of the light is found. The depth at which the light intensity is reduced to 1 per cent of its surface value is of particular importance because it represents the approximate lower limit for plants as will be discussed later. In general the table and the representative curves profound variation in the intensity of light available for organisms at increasing depths in different natural waters.

Table 1.2 Transparency of water to central part of visible spectrum (50000-60000Å)

	Extinction per meter (%)	*k**	*Depth for 1% of Surface Light*	*Secchi Disc Depth*
Distilled water	3.8	0.039	118 m	44 m
Caribbean Sea	4.1	0.041	110	41
Continental slope	7.2	0.072	60	24
Gulf of Maine	10.0	0.10	42	17
Woods Hole Harbor	26.0	0.30	16	6
Crater Lake, Oregon	6	0.06	77	28
Crystal Lake, Wis.	15	0.16	28	11
Trout Lake, Wis	33	0.40	11	4
Midge Lake, Wis.	78	1.5	3	1

* Extinction coefficient, *k*.

$$\frac{I}{I_0} = e^{-kL}$$

where I_0 = initial intensity.

I = intensity at depth, L, in meters.

$e = 2.7$.

Another way in which light is changed by water is in its angular distribution. The direct beam of light from the sun is bent by refraction toward the perpendicular at the water surface. Since waves or ripples almost always exist in natural situations, the light from the sun tends to be broken up to a considerable extent in passing through the surface. Within the water itself further diffusion is produced by scattering. When larger amounts of suspended materials are present, scattering proceeds at a very rapid rate, with

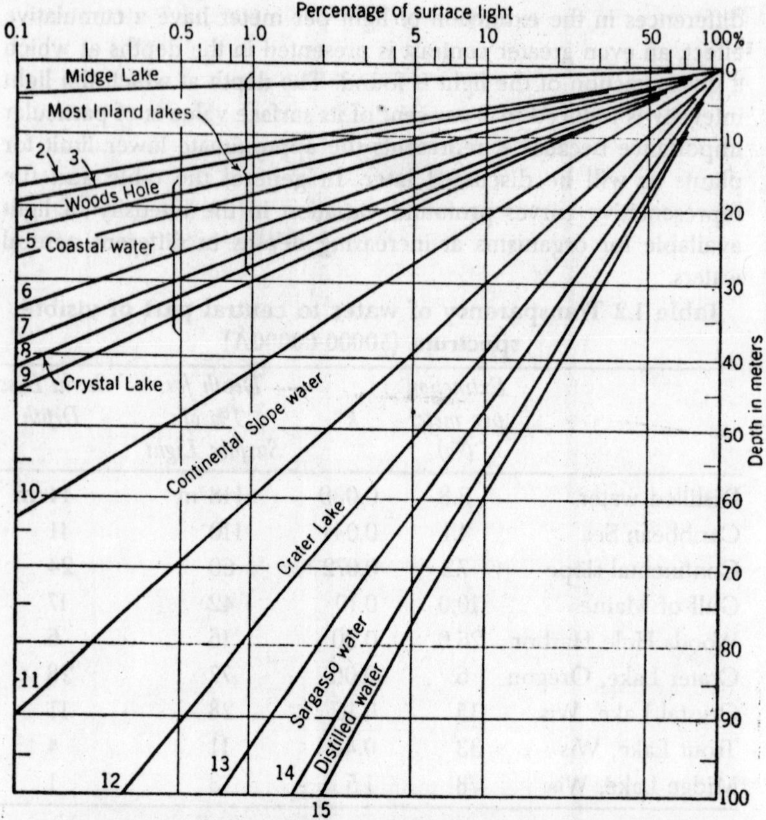

Fig. 1.7. Comparison of the transparencies of representative natural waters. Curves based on the average extinction rate for the yellow-green (5000–6000A) component of daylight.

the result that measurable light is passing in every direction. In the upper strata the light passing in the direction of the refracted rays from the sun is the strongest. With increasing depth the direction of strongest light intensity tends to move toward the vertical, and a limiting pattern of diffusion is established. Thus, although a variable degree of diffusion occurs, the directional character of the light in the water is never entirely lost.

Changes in transparency

Parts of the ocean and of lakes that are well stirred exhibit a uniform transparency from the surface downward, but, when strong density gradients exist, considerable changes in transparency may occur with depth. Furthermore, the transparency of coastal marine areas and of inland water bodies often changes profoundly from season to season, owing to differences in amount of stirring, in the discharge of muddy rivers, and in the growth of plankton. Such variations in transparency are added to the seasonal changes in the intensity of the light received at the surface and in the length of the day. The combined effect of these influences results in surprisingly great fluctuations in the light factor at subsurface levels. Observations made off the coast of Massachusetts illustrate the magnitude of the foregoing seasonal influences.

In this region the maximum hourly intensity of light received at the surface during the summer averaged about twice that received during the winter period. The average total solar radiation received per day in summer was about four times as great as that received during the winter. Significant departures from the average for periods up to several weeks occurred, and such variations are undoubtedly important in causing changes in the response to the light factor from year to year. Values for individual days may also depart markedly from the average. During the investigation referred to above the maximum light received on the brightest day of summer was forty times greater than that received during the dullest day in winter.

In the course of the same year the extinction coefficient of the water varied four-fold. Since the effect of extinction is cumulative, the illumination in the subsurface layers varies more because of changes in transparency than because of seasonal changes in radiation reaching the surface. By combining the two effects it was found that for an organism living at a depth of only 80 m the maximum hourly light intensity was 7000 times greater and total daily radiation was 10,000 times greater in May when high incident light was combined with high transparency, than in December when both incident illumination and transparency were low.

The magnitude of seasonal charges in illumination in the aquatic environment are therefore frequently very much, greater than in the terrestrial environment. Any water body that displays a much higher transparency in winter than in summer will have less light available during the summer in its deeper layers and thus suffer a reversal of the usual seasonal change in the light factor the foregoing instances are sufficient to show that in the aquatic environment light becomes profoundly altered quantitatively and qualitatively and that its changes may go far beyond those experienced by plants and, animals on land.

Orientation

Light often plays a significant role in orienting the growth, or locomotion of plants and animals. Since orientation to, light is often associated with reactions to other factors in the environment, such as gravity, we shall first consider the subject in general terms. Orientation is brought about either by the differential growth or movement of parts of the organism or by a change in the direction of locomotion of the whole' organism. The question of why plants and animals grow or' move in the directions they do has occupied the attention of investigators for a long time.

Many of the theories and terms used have been in conflict, and for a more extended discussion reference should be made to Fraenkel and Gunn (1940) and to Griffin (1953). The term *tropism* is best used for orientation by growth or turgor movements as exhibited by sessile forms. These forms are usually plants, but essentially the same type of orientation is exhibited by many sessile animals such as the hydroids. If the orientation is to gravity, the term *geotropism* is used. If it is to light, the growth movement is refereed to as *phototropism* and other prefixes are used for other orienting factors. On the other band, the orientation of the

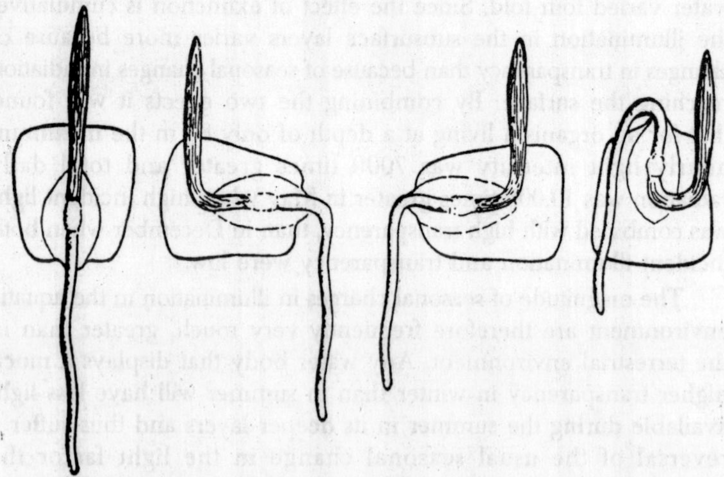

Fig. 1.8. Negative geotropism of shoots and positive geotropism of roots of four kernels of maize that have germinated in different positions.

locomotion of motile organisms is best referred to as a *taxis*, although it is sometimes also called a tropism. Here the forms involved are usually animals although motile plants such as the green flagellates and motile Plant gametes or zoospores are included.

Applying suitable prefixes; we obtain the terms *geotaxis*, *phototaxis*, etc., for orientation by the corresponding factors. Orientation in the direction of an orienting force is referred to as a positive tropism or taxis; orientation in the opposition direction is a negative tropism or taxis. Occasionally a transverse tropism or taxis is displayed. Control of speed of locomotion by the intensity of the factor is termed a *kinesis.* Gravity never changes significantly in intensity, but the speed of swimming, flying or creeping is influenced by alterations in the strength' of other factors; tile common occurrence of photokinesis has already' been mentioned. The position of the main axis of the body is the primary orientation

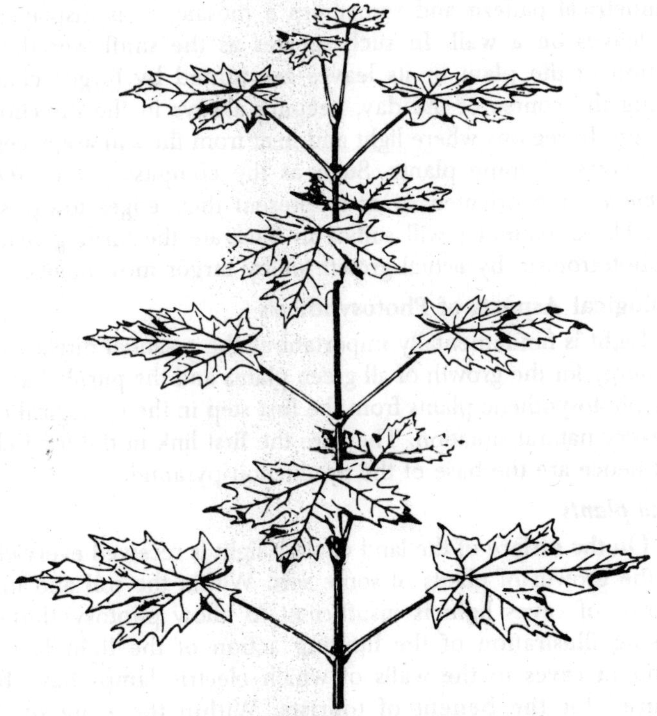

Fig. 1.9. Alternating position and horizontal orientation of leaves on shoot of Norway maple, resulting in maximum exposure to sunlight.

of the plant or animal. This is usually determined by gravity, as is seen in the upward growth (negative geotropism) of the shoot and the downward growth (positive geotropism) of the root of a plant seedling. In the aquatic environment the buoyant action of water often reduces the effect of gravity so that primary orientation is to light or to current.

When the source of illumination is from the side the primary orientation of, a land plant will be a compromise between the influence of gravity and that of light. The primary orientation of motile organisms is also usually to gravity, and a walking or a flying animal continually adjusts its position to it. Superimposed upon the primary orientation, many secondary responses to orienting influences take place. The tips of green plants grow toward the light and hence exhibit positive phototropism. The leaves are oriented generally at right angles to the incident radiation, thus receiving maximum illumination and they sometimes form a symmetrical pattern and sometimes a mosaic, as is displayed by ivy leaves on a wall. In such species as the sunflower the top portion of the plant or its leaves are turned by turgor changes during the course of the day, keeping always in the direction of the sun. In regions where light and heat from the sun are excessive the leaves of some plants. Such as the compass plant *silphium laciniatum*, are oriented so as to present their edges towards the sun. These examples will suffice to illustrate the turning reaction of phototropism by actual growth or by turgor movements.

Ecological Aspects of Photosynthesis

Light is fundamentally important as the essential direct source of energy for the growth of all green plants and the purple bacteria. The photosynthetic plants from the first step in the ecological cycle in every natural situation. They are the first link in the food chain and hence are the base of the production pyramid.

Land plants

On the surface of the land enough light is received everywhere for the growth of plants of some sort. Within the soil and in the interior of caves light is insufficient to allow photosynthesis. A striking illustration of the limiting action of the light factor is found in caves to the walls of which electric lamps have been secured for the benefit of tourists. Within the cone of light immediately around each lamp mosses have grown from spores brought in by air currents. These "islands" of plant growth present a sharp contrast to the complete lack of vegetation in the remainder of the cave. Although illumination is generally sufficient for photosynthesis over the entire surface of the globe, the local distribution of individual species of plants is definitely influenced by differences in the availability of light. Species that can grow in shady places are termed *tolerant*, and, although the degree of tolerance is affected by soil moisture, temperature, and other factors, the illumination is the principal controlling influence.

Plants that require strong illumination and will not survive or develop in reduced light are referred to as *intolerant* species. Trees such as the spruce, hemlock, beech, and sugar ,'maple, shrubs, such as spicebush, and herbs, such as bloodroot, can grow in deep shade and are examples of tolerant species. The sugar maple can photosynthesize adequately in situations where the illumination is reduced to less than 2 per cent of full sunlight. Birch trees, poplars, willows, and several species of pine, as well as many shrubs and herbs such as sumac, bluestem grasses, and sunflowers are intolerant of shade. The ponderosa pine, for example, requires a light intensity equal to at least 25 per cent of full sunlight. Intolerant species cannot develop in the shade of a dense stand of their own or other species.

Even among somewhat tolerant species the first plants to become established in an open area will grow to better advantage than individuals subsequently developing around them because of the competition for light and other needs. Differential growth often results in "pyramiding" in the development of such a group of plants. As we have seen in an earlier section, the light on the floor of the temperate deciduous forest become seriously reduced when the trees leaf out. Many of the smaller plants are adapted to a period of rapid growth in the early spring while light on the forest floor is sufficiently intense. In the marginal rainforest of Panama a dense undergrowth is possible because during the dry season the leaves fall from the trees and allow light to penetrate.

In the equatorial rainforest of Columbia, on the other hand, trees shed their leaves individually and a continuous canopy is maintained, with the result that little vegetation can grow on the forest floor. In such rainforests many smaller plants have gained access to sufficient illumination by the evolution of the epiphytic habit, that is, by growing in the crotches or on the horizontal branches of the trees. The vertical distribution of the many species of bromeliads in the forest trees of Trinidad reveals their varying degrees of tolerance. One group of species grows only near the tree tops in situations fully exposed to the sun, a second group is found at intermediate levels in partial sunlight, and a shade-tolerant third group inhabits the lower branches of the forest where the illumination is quite inadequate for the other groups.

Aquatic plants

Reduction of light presents even more serious problems in the aquatic environment. As we have seen illumination diminishes rapidly with depth even in clear water and becomes changed in spectral composition and in other respects. Plants attached to the bottom in the marine environment consist principally of algae with a few species of vascular plants such as the eel grass. In fresh water vascular plants as well as algae are well represented in the submerged vegetation. Early investigators reported an apparent colour zonation in the depth of occurrence of attached plants. In

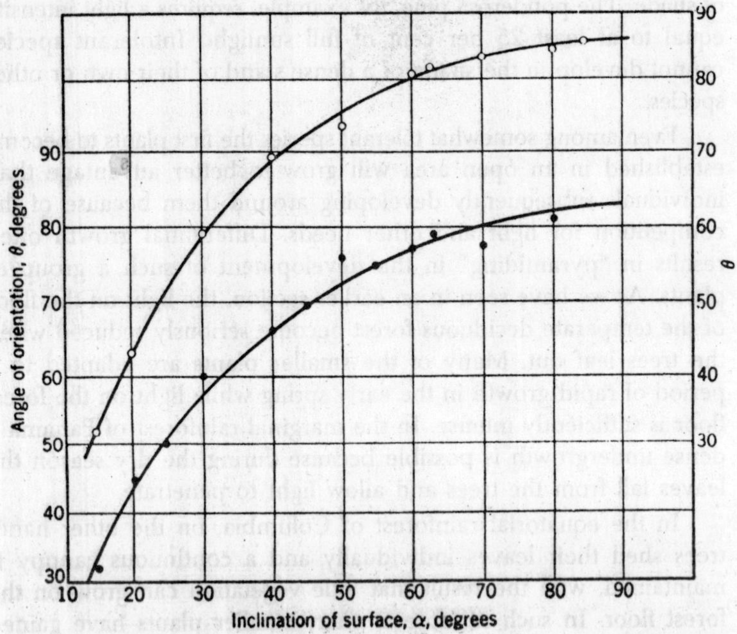

Fig. 1.10. Relation between the angle of inclination of plane (α) and the mean angle of orientation (θ) of caterpillar upon it. Open circles are means for one individual; black circles are means for all individuals tested. The curve is that for Δθ/Δ log sin α=constant.

Puget Sound, for example, the green algae are generally the most abundant in shallow water, brown algae dominate the zone from 5 to 20 m, and red algae are usually most numerous in depths of 10 to 30 m. These colour types of algae often occur at depths where the complementary colour of the penetrating daylight predominates-red algae, for example, tend to be abundant in deep water where the blue or green component of daylight is the strongest.

It was formerly believed that the predominating colour of the light controlled the depth distribution of the colour types of algae because the plants would absorb light of complementary colour more efficiently. This generality was referred to as *chromatic adaptation*. Many exceptions occur, however, and we now know that the wavelength of light does not control in any precise way the depth of distribution of algae on the basis of their colour. Furthermore, the pigment of algae living in weak light deep in the water is sufficiently thick to absorb all the light incident upon it, regardless of the part of the spectrum in which it occurs.

Evidence exists that some of the light energy absorbed by pigments other than chlorophyll can be transferred to the chlorophyll present and thus enhance the ability of certain algae to live in weak blue-green light. When algae are grown, in the laboratory in light of different colours, they often tend to change colour, and this response, has also been referred to as chromatic adaptation. Experiments have shown that individual plants photosynthesize most efficiently and grow best in that colour and in 'that intensity in which they have been living. In the course of time fixed algae can adapt themselves to habitats with illumination of widely different conditions of colour and intensity. In view of the great range in the transparency of natural waters, it is not surprising that the maximum depth at which attached plants can live varies widely in different situations.

In general benthic plants will not grow at depths at which the light intensity is less than 0.3 per cent of the surface value. A glance at a chart of the oceans or of a typical lake reveals that only in a very narrow fringe around the margin is the water sufficiently shallow for enough light to reach the bottom for the growth of plants. We are forced to the realization that in any deep body of water far more organic matter is synthesized by the phytoplankton present everywhere in the surface layers than by the benthic plants limited to the littoral zone. Photosynthesis, the first step in the ecological cycle of the sea upon which all marine life ultimately depends, is carried forward primarily by the minute diatoms and other microscopic planktonic plants. The growth of the phytoplankton is limited not only by the fact that vertical movements of the water subject the plants to continually changing light conditions. The greatest depth at which phytoplankton can grow successfully in any body of water may be ascertained by suspending bottles containing samples of the plant population at various levels beneath the surface.

Table 1.3. Maximum depth in meters for Growth of attached plants.

Crater Lake, Ore	120
Crystal Lake, Wis.	20
Trout Lake, Wis.	12
Mediterranean Sea	160
Challenger Bank	100
Of Iceland	50
Puget Sound	30
Baltic Sea	20
Off Cape Cod	10

Photosynthesis going on within the bottle adds oxygen to the water, and respiration taking place simultaneously removes oxygen from the water. The rate of respiration alone may be measured by placing at each level a blackened bottle from which all light is excluded. The sum of the gain in oxygen in the clear bottles and the loss in the dark bottles gives a measure of total photosynthesis. The increment of oxygen in the clear bottles is the amount produced by photosynthesis minus the amount consumed in respiration and represents the actual increase in energy content or growth of the plants. With the diminution of light in any environment photosynthesis is reduced but respiration remains approximately the same, provided, of course, that temperature and other factors are essentially unchanged.

For each species a point is reached in the reduction of the illumination at which the rate of photosynthesis is just equal to the rate of respiration; this is known as the compensation point or better as the *compensation intensity*. At light intensities below this value photosynthesis may still go on but the plant is fighting a losing battle: it cannot survive indefinitely under these conditions because the loss of energy due to the catabolic processes represented by respiration exceed the gain in energy brought about by the anabolic process of photosynthesis.

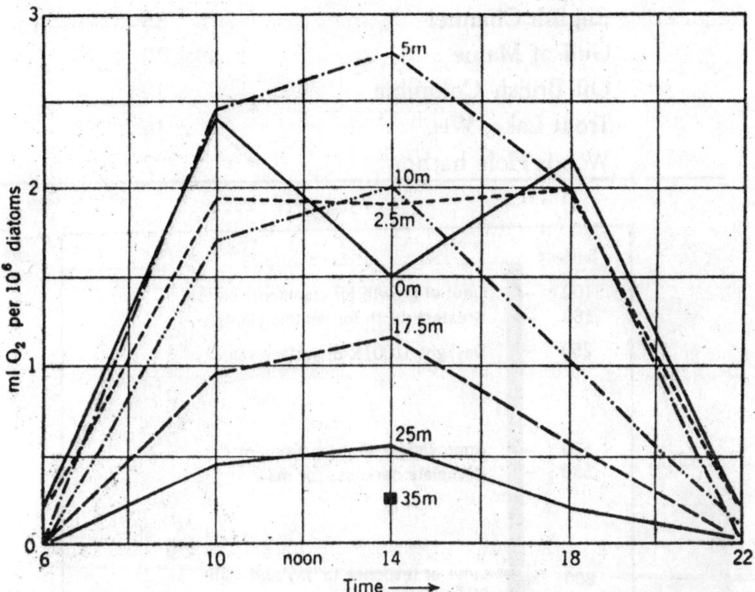

Fig. 1.11. Photosynthesis of a species of phytoplankton (Coscinodiscus) at the indicated depths during the course of the day off Stoke Point, England.

With further light reduction a minimum intensity is reached below which no photosynthesis at all can take place. In the aquatic environment the level in the water at which the compensation intensity is found is called the *compensation depth*. Sample values for phytoplankton based on measurements in the middle of the day are presented in Table 1.4. For phytoplankton in general the compensation intensity, has been found to be about 1 per cent of the value-of full sunlight at the surface. The change in the amount of photosynthesis at various depths during the course of the day is plotted for a type of diatom representative of the marine plankton. It will be observed that during the noon hours the illumination close to the surface was excessive and depressed photosynthesis. The highest rate of oxygen production occurred at 5 m in the middle of the day. No significant amount of photosynthesis occurred at any depth before 0600 in the morning or after 2200 and none below 35 m at 1400.

Table 1.4. Compensation depths in meters for phytoplankton (for midday periods)

Sargasso Sea	100
English Channel	35
Gulf of Maine	30
Off British Columbia	19
Trout Lake, Wis.	16
Woods Hole harbor	7

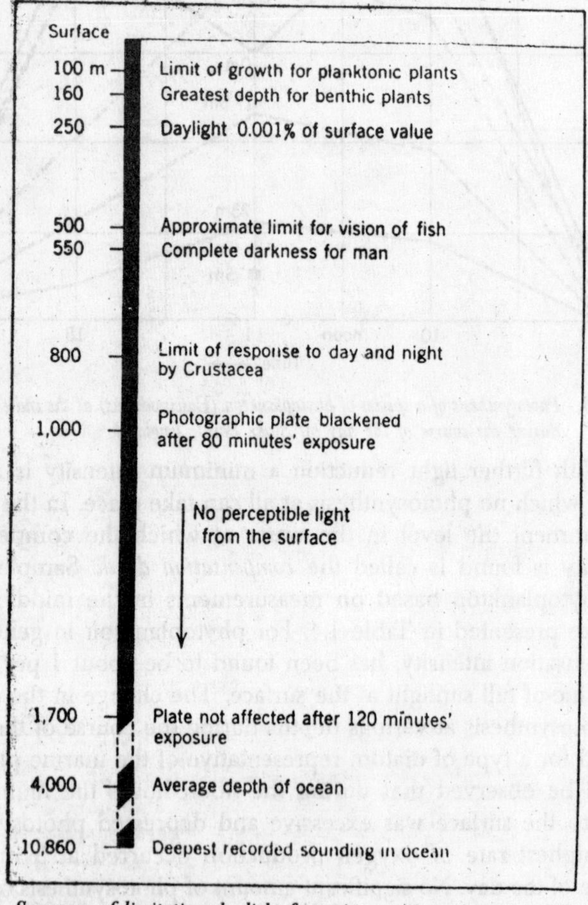

Fig. 1.12. Summary of limitations by light factor in aquatic environment. Values are for clearest water.

If a plant is to grow, its photosynthesis during the day must build up enough organic matter to more than make up for the material lost by respiration not only during the day but also during the night. In other words, the crucial value for the continued existence of the plant is the compensation depth for the 24-hour period. This will obviously occur at a shallower depth than the values reported for experiments limited to the middle or the day. The compensation depth for the complete day ranges between 20 and 30 m during the summer in clear coastal water of the temperate oceans. In the winter and in less transparent water the compensation depth occurs correspondingly nearer the surface. No constructive growth is possible for diatoms or other pelagic plants below these levels.

When we recall that the average depth of the ocean is more than 4000 m, it is apparent that the zone in which productive growth of the crucially important phytoplankton can take place is a relative skimming of the surface. The photosynthesis in the upper few meters of the ocean and of takes accounts for the main portion of the initial production of organic matter for the whole breadth, length, and depth of the water. Here are indicated the maximum depths for the growth of phytoplankton and slightly deeper, of benthic plants. Since the mean illumination necessary fur the vision of aquatic animals is so very much smaller, the depth limitation is at a much greater level. Animals can respond to the differences between day and night at somewhat greater depths. Below this level no perceptible light from the surface penetrates and the water is completely dark except for light provided by the luminescent organs of deep sea animals. Terms used in deep bodies of water for zones based on the light factor are as follows:

Euphotic Zone : Sufficient light for photosynthesis.

Disphotic Zone : Insufficient light for photosynthesis but sufficient light for animal responses.

Aphotic Zone : No light of biological significance from the surface.

The depth limits of these zones differ widely according to transparency. The above discussion has revealed the many important roles played by light in the world of life. Although there is generally sufficient light on land for plants and animals, the special circumstances of this factor exert a wide control over the activities of many terrestrial organisms. In the aquatic environment light is even more crucial since, in addition to its various periodicities, its rapid changes with depth impose serious limitations on the lives of both animals and plants.

Central Importance of Light to the Physiology of Plants

Although plants do not have a total monopoly on photophysiology, no other group of organisms has the same range of responses to light: photosynthesis and other light-stimulated, chemical transformations, photoperiodism, and phototropism, to name the most prominent. Moreover, studies with plants have laid the groundwork for photophysiological research, generally and, in some cases, have contributed to or simulated understanding in photochemistry and spectroscopy. The analysis of a photophysiological phenomenon consists essentially in obtaining answers to two photochemical questions together with some attendant ones. What is the absorbing pigment? And what are the reactions of the excited pigment? The theories of photochemistry and spectroscopy can become quite complicated, especially if we begin on worry about energy levels in excited states and the mechanisms of photochemical reactions. But we shall see that the limited questions of identity and reaction products may be answerable with the evidence of action spectra, absorption or difference spectra, spectra of light emission, and electron spin resonance.

Photochemistry

The study of known photochemical processes permits us to anticipate the kinds of processes we shall encounter in photosynthesis and other biological responses to light.

Electronic Transition in Atoms, Molecules, and Crystals

Light can be drought of as traveling in packets called photons. Photons are often designated as hv, which is at once a symbol and a representation of their energy. Every time an ultraviolet or visible photon is absorbed the absorbing material undergoes an *electronic transition.* The simplest case of an electronic transition occurs in a hydrogen atom. The single hydrogen electron is exicited to a new energy level; the *entire* energy of the photon is converted into additional kinetic energy of the electron. Just as a violin string may vibrate only in a fundamental tone and higher harmonics, the hydrogen atom is stable only in certain definite energy states, so photons of certain definite energies, and only these energies, are absorbed by hydrogen atoms.

It is precisely as if the hydrogen atoms resonated at certain critical frequencies (or wavelengths) of light. If, with a spectograph, we were to observe white light that had passed through hydrogen atoms, we should see absorption at the wavelengths corresponding to transitions of the hydrogen *atoms*. The result is a series of absorption lines (Fraunhofer lines); atoms such as hydrogen are thus said to exhibit "line spectra." The general method of wave mechanics which treats orbital electrons as simple harmonic oscillators, is fairly successful even with complicated atoms such as sodium and roil.

In contrast, the transitions of even the simplest molecules are enormously more difficult to dissect, but approximations are possible by starting from the premise that the most stable states for chemical bonds correspond to a minimum energy mode in which an electron wave can oscillate between two nuclei. The same kind of electronic transitions occurs in molecules as in atoms and is responsible for the absorption of visible and ultraviolet light by molecules. Wave mechanics also provides a rational basis for the phenomena of "allowedness" and "forbiddenness," i.e., the complete or partial exclusion of certain types of transitions. Similar considerations apply to the energy levels of crystals, in which there can exist free electrons, i.e., electron waves oscillating within conduction bands or whole regions of the crystal.

Other Transitions

We have been talking about electronic transitions. There are other kinds of transitions in polyatomic molecules and crystals: (1) vibrational transitions, where the energies are contained in elastic vibration among the nuclei of the molecule or crystal, and (2) rotational transitions, where the energies are associated with the twisting of model along the axis of the bond joining them. The energies of pure vibrational and rotational spectra are weak, and so the photons that can induce such transitions are restricted to the infrared. (Indeed, infrared absorption spectra are specific indicators of molecular structure for the simple reason that they correspond to the vibrational and rotational modes possible for the molecule.) The photon energies for these different kinds of transitions are easily calculated. The energy of a photon is equal to the product of Planck's constant and the frequency of light

$$E = h\nu \quad \text{or} \quad E = \frac{hc}{\lambda}$$

where c is the velocity of light and λ is wavelength. The energy of a photon thus may be determined from the value of Planck's constant and the wavelength or frequency of light. When the wavelength is expressed in mμ. (or nm), the energy in electron volts is

$$E_{ev} = \frac{1,235}{\lambda_{m\mu}} \quad \text{...(2)}$$

The approximate energy relations among the different kinds of transitions are:

Transition	*Effective Wavelengths, mμ*	*Energy electron volts*	*calories*
Electronic	100–1,000	1–10	300,000–30,000
Vibrational	1,000–10,000	0.1–1	30,000–3,000
Rotational	10,000–100,000	0.01–0.1	3,000–300

In the case of molecules, we usually observe mixed spectra. In these cases, the molecule exists in one of several possible vibrational states associated with the ground electronic state. The molecule may be raised to any of several possible vibrational states associated with an excited electron state. Instead of exhibiting the relatively simple line spectra of atoms, molecules show extremely complex spectra. In the gaseous state, the spectra consist of bands of relatively diffuse lines. In the condensed phase, the multitude of vibrational and rotational states smears into a single blurred peak for each electronic transition. The ultraviolet and visible absorption spectra which have become such a commonplace in chemistry and biochemistry therefore have their physical basis in electronic transitions, greatly modified by vibrational and rotational transitions.

Transitions may occur to the first, second, or higher excited states, For example, with chlorophyll *a,* light of 663 mμ. which corresponds to 41 kcal is sufficient to excite the molecule to the first excited state. Light of 410 mμ (65 kcal) exciteş chlorophyll to the second excited state. Once in the excited state a molecule may be excited still further. Again in the case of chlorophyll *a,* it has been possible with tremendous light intensities to convert most of a given sample to another excited state (in this case the *triplet state*).

It remains in this state long enough that one may irradiate the solutions again, this time with monochromatic light and this time observe transitions to the second *triplet* state. Unlike electronic transitions, vibrational and rotational states are functions of the temperature. We may therefore expect that different absorption spectra might be observed at different temperatures. In fact, absorption spectra at extremely low temperatures show greatly sharpened peaks.

Extinction Coefficient and Oscillator Strength

If we are to characterize a particular electronic transition for a polyatomic molecule in the condensed phase, we have two quantities that we can measure: (1) the wavelength of the absorption peak and (2) the intensity of the absorption peak. The most familiar measure of absorption intensity is the extinction coefficient, simply defined in the Lambert-Beer equation.

$$\log_{10}\frac{I_0}{I} = \varepsilon cd \qquad ...(3)$$

Where I_0 and I are initial and final intensities of light passing through a sample, c is the concentration of the sample in moles per liter, and *d* is the light path ill centimeters; the extinction coefficient, ε, then has the units of liter $mole^{-1}$ $centimeter^{-1}$. A more fundamental quantity is *oscillator strength*, which the efficiency with which an oscillator (electron) resonates with the field (light). An ideal oscillator (e.g., a hydrogen atom) absorbs light in an allowed transition with 100 per cent efficiency. We say it has an oscilla or strength $f_A = 1$. Real molecular oscillators absorb light over a band of wavelengths,) that we must integrate (see Equation 4). The integrated oscillator strength f_A is

$$\bar{f}_A = \int_{\lambda_1}^{\lambda_2} f_A(\lambda)d\lambda \qquad ...(4)$$

There is a convenient relation between oscillator strength, adsorption coefficient (Equation 3) and the index of refraction the medium, ***n***.

$$\bar{f}_A = 4.319\times10^{-9n}\int_{\lambda_1}^{\lambda_2} f(\lambda)d\lambda \qquad ...(5)$$

Oscillator strength sums the electronic transition over all of the vibrational and rotational substates. This quantity represents, therefore, the total probability or allowedness of the electronic transition occurring. Since the probability of the transition occurring is the same in both directions, the oscillator strength is also related (as is the absorption coefficient) to the reciprocal of the mean lifetime of the excited state. Nearly ideal organic oscillators have strengths of 1 corresponding to absorption coefficients of about 10^6 liters mole^{-1} cm^{-1} and mean lifetimes of about 19^{-9} sec.

Futes of Excited Substances

When a substance absorbs a photon in the visible or ultraviolet, it undergoes an electronic transition and becomes excited.

$$A + hv \rightarrow \text{A} \qquad ...(6)$$

The excited substance may then undergo singly or in some cases in sequence:

1. Ionization.
2. A chemical reaction.
3. Fluorescence.
4. A long-lived or metastable state, such as the triplet state.
5. Isomerization (if a molecule).
6. Dissociation (if a molecule).
7. Transference of the energy of excitation to another atom or molecule at a small distance; this is called energy transfer; excition migration, or resonance transfer.
8. Conversion of all or part of its energy of excitation to kinetic energy through what is called internal quenching or a radiationless transition.

Since all of these processes may possibly occur in intact plants, we shall examine them more closely.

Ionization

From the energy diagram of the hydrogen atom, we saw that a photon of 13.60 ev (or 91.1mµ) was sufficient to move the hydrogen electron completely free of the nucleus. This is of course an ionization. Organic molecules may lose electrons at much lower energy levels: radiation of wavelengths shorter than about 300 mµ, frequently called "ionizing radiation," is famous for producing genetic mutations. Ionized species may subsequently enter into chemical reactions. Moreover, the acid dissociation constants of molecules may be enormously altered by excitation. The P_k of β-naphthol is 9.5, whereas the P_k^* is 2.5. Thus β-naphthol in the excited state is a moderately strong acid !

Chemical reactions

Excited species may enter directly into chemical reactions. The reaction may occur simply because that energy of activation supplies the energy of activation of the reaction; that is, the energy of excitation "drives" the reaction, producing a net decrease in free energy. A reaction may also occur because the orbitals of the excited state labilize a bond ,greatly increasing the possibility of orbital overlap with another reactant. The last-named mechanism is in effect a decreased energy of activation for the reaction. Because of the extremely short lifetimes of singlet states, the probability is low for their experiencing a favourable collision. Efficient bimolecular reactions involving singlet excitation would require extremely favourable steric factors. Reactions with molecules in forbidden states are quite another matter for the very reason that they are long lived. There have been frequent suggestions that triplet states mediate a wide variety of reactions.

Fluorescence

Atoms in the gas phase may emit the simplest and of fluorescence, namely reemission of the absorbed photon. This is called resonance fluorescence. In polyatomic molecules and in condensed phases, fluorescence occurs when the species is excited to a higher singlet state. Part of the enemy is lost by internal quenching (alternative 7 above), bringing the species down to the first singlet state. The remaining energy is then reemitted as a photon. Since the probability for a transition is the same in either direction and the average lifetime of a single state (assuming an allowed transition to the ground state) is not much longer than 10^{-8} sec, fluorescence usually occurs within 10^{-9} sec. Longer intervals (up to 10^{-9} see) have been observed when exciton migration intervenes between excitation and fluorescence.

Long-lived and metastable stares

Although transition from the ground state to forbidden states in highly improbable, it may be quite probable for a molecule to pass from an excited singlet state to a forbidden one. Photons may then leak slowly from the excited species resulting in *phosphorescence.* A typical forbidden state is one with two unpaired electrons. From the observation that, under the influence of an external magnetic field, atoms with unpaired electrons can exist in three excited states with closely similar energy levels, spectroscopists named such little constellations "triplet states." The term has been used loosely for any excited species which emits light over a long interval, even when the triplet energy splitting has not been demonstrated. This imprecision is unfortunate, especially in biological systems where highly ordered. Near crystalline arrays may occur in which other long-lived or metastable states are possible.

Isomerization

An extremely interesting example of photochemical isomerization is the interconversion of maleic to fumaric acid. Although these two acids have quite distinct structures, corresponding to the cis and trans configurations around a carbon-carbon double bond, the excited states of the two are nearly identical so it is possible for an excited fumaric molecule to return to the maleic ground state and *vice versa.* Photoisomerizations may be of great importance in biology.

Dissociation

If the energy of excitation is transformed into vibrational modes, it is quite possible for the new vibrational level to be sufficiently great to break the chemical bond. The "envelope" for vibrational energies is apt to be much more shallow in the excited than in the ground state and displaced toward larger distances between the nuclei, so that after an electronic excitation from a stable vibrational state, a molecule can find itself in an unstable vibrational state and the molecule dissociates. The resulting fragments are free radicals, which are typically extremely reactive. Many light-catalyzed polymerizations and other chain reactions are initiated by photochemically generated free radicals. A photodissociation is also thought to be involved in the release by light of cytochrome oxidase from carbon monoxide inhibition.

Quenching

Even though increased electronic energy resulting from excitation may labilize a bond sufficiently to permit a chemical reaction, the more usual event is that the excitation energy is converted to vibrational energy among all the bonds of the molecule. Through the principle of equal partition of energy, ultimately all of the excitation energy is converted to heat. The initial conversion of electronic to vibrational energy is called internal conversion or quenching.

In the liquid phase and at ordinary temperatures, collisions frequently enhance quenching, so that most photochemical reactions, including fluorescence and phosphorescence, are strongly quenched in liquid phases at ordinary temperatures. Considering the enormous variety of ways in which molecules can lose their excitation energies without entering into a chemical reaction, energy conserving photochemical reactions such as photosynthesis seem highly improbable!

Energy transfer

Exactly as a plucked violin string will induce resonant vibrations in an adjacent string tuned to the same frequency, an excited electron in a molecule will induce an electronic transition in a nearby molecule. More than that, if the resonance energy of the neighbouring molecule is some what less than that of the excited molecule but still sufficiently near that their absorption bands overlap a transference of all of the excitation energy must occur. In this way excitation energy can migrate across many molecules in a closely packed array. The phenomenon is called energy transfer, exciton migration, and resonance transfer. According to one model, the efficiency of energy transfer should decrease with the sixth power of the distance separating the molecules.

Photochemical Evidence

Let us suppose that we come across a biological process in which light appears to be involved. The process might be oxygen evolution, movement, change in pigmentation flowering, luminescence, release from CO inhibition, etc. What questions might we appropriately ask about such a system? And by what criteria should we judge answers to these questions? Questions and the kinds of answers to them are the subjects of this section. The questions we ask are simple enough What pigment or pigments absorb light? Where are the pigments? What is the excited state, What is the fate of the excited pigment molecules. That is, what transformations or reactions does the pigment experience ? How many quanta are required for a unit step in the process? Is there transfer of ıadiant energy in the system and if so how does transfer occur?

After the strictly photochemical questions, there are questions that grade rapidly into biochemistry. For example, is the process strictly photochemical, or are there accompanying "dark" reactions, which might be required to regenerate the pigment? How does the product of the light reaction influence the physiological process ? The kinds of evidence required to answer these questions are all based squarely on the principles of photochemistry.

The Bunsen-Roscoe Law

Our first step in appraising a light-influenced process is to determine if the response is proportional to the amount of light absorbed.

$$R = kIt \quad ...(7)$$

where R may be taken as the extent of a process, I is the light intensity, and *t* is time. This statement is known as the Bunsen-Roscoe law. It follows directly from the notion of excitation

$$A + hv \rightarrow A^* \rightarrow P$$

We can expect that, since every A* has an equal probability of resulting in P, the rate or extent of the reaction should be proportional to A*; the light absorbed is proportional to the concentration of pigment A and the amount of light, which is clearly the product of I and *t*. Nonconformity with the Bunsen-Roscoe law means that 'either A varies over the time intervals chosen, A varies with I, or the extent of the reaction is not proportional to the immediate photochemical product P.

Variation of A with I can occur in the case of optical rarely encountered in biological systems, In the scheme

$$x \rightarrow A + hv \rightarrow A^* \rightarrow P$$

variation of A with time can mean that the rate of formation and/ or regeneration of A is slower than (or of the same order of magnitude as) the rate of excitation (absorption).

$$\frac{d(x \rightarrow A)}{dt} + \frac{d(A^* \rightarrow A)}{dt} \leq \frac{d(A \rightarrow A^*)}{dt}$$

We may use flashing light tests to tell us if a "dark" reaction is required for the light-influenced process. By varying the dark interval between flashes, we may learn something about limiting rates of the dark reaction. The effects of variations in the intensity of duration of the flashes give corresponding information about the relative amount of absorbing pigment.

Action Spectra

The formation of the immediate photochemical product P is proportional to A* and therefore also to the absorption of light by A. From this rather obvious statement, it follows that if a chemical or physiological response to light, R, is proportional to P, then R is proportional to A (Equation 8) for all competent transitions.

$$R(\lambda) = kA\ (\lambda) \quad ...(8)$$

A graph of R(λ) is said to be an *action spectrum*. Equation 8 tells us that the action spectrum should have the same shape as the absorption spectrum of the absorbing pigment. Equation 8 is subject to certain obvious restrictions: that R is in fact proportional to P, that other pig:nents do not shade or mask A, that energy transfer does not occur, etc. Stated differently, when the action spectrum does not have the same shape as the absorption spectrum of a pigment, determined on other grounds to be the *absorbing pigment*, then one or more of the above restrictive conditions must obtain, For further discussion of some of the pitfalls of action spectra. In general, the experimenter must ensure that the Bunsen Roscoe law is obeyed when measuring action spectra.

Absorption Spectra

The simple requirements that (except in lasers) there must be an absorbing pigment in order for light to produce an effect and that A→A*, mean that photochemical processes may result in changes in the pigments of the system; *e.g.*, disappearance of A. This is obviously not a necessity; A* may be vanishingly small compared with A, and A may be rapidly regenerated. If absorption changes are observed, we may imagine them to be due to either disappearance of A or appearance of either A* or P. The occurrence of absorption changes due to the appearance of products may be further tested by studying the sequence of absorption changes following excitation. The changes observed *in vivo* may correspond to absorption spectra or difference spectra of the participating pigments.

Light Emission

If the reader has gained the impression that measurements of light absorption lire powerful and versatile instruments in the physiological armory, he should immediately set down measurements of light emission as an even more sensitive and almost as versatile means of inspecting physiological processes. The reason is purely technical; light absorption requires the comparison of the intensity of one light beam with another and is dependent on the stability of detectors and amplifiers. In contrast, the emission of light may be detected at the highest sensitivities of photocells and photomultipliers where the "control" is simply the dark current of the detector.

The sensitivity for measuring fluorescence can be enhanced even further by modulating the exciting radiation and locking the detector to the same frequency. When an excited molecule emits light, we may measure the rate of decay of light following removal of the exciting radiation. From the lifetime of the excited state, we can determine whether the light is fluorescence or phosphorescence. From measurements of depolarization we can also inter whether exciton migration intervened between excitation and emission. We may also determine the *emission spectrum*, from which the identity of the excited species may be deduced.

And, finally, we may determine the *excitation spectrum* or *action spectrum* of fluorescence, which has the same shape as the absorption spectrum. A fluorescence emission spectrum resembles a long wavelength mirror image of the singlet absorption band. This occurs because internuclear distances are greater in the excited state than in the ground state. Consequently an excited molecule is apt to return to a higher vibrational state than the one it occupied in the ground state. The energies of the photons emitted will be less energetic, and hence of longer wavelengths, than those absorbed by the molecule in the ground state.

If the case of exciton migration, the absorbing species will very likely be different from the emitting species. This is typically the case in chlorophyll fluorescence in green plants. A caveat: emitted light energy of fluorescence and phosphorescence is of course not available for photochemistry. The molecule emitting fluorescence or phosphorescence may be the same as those involved in a photobiological process, but it may also represent a diversion from the photochemistry of the physiological response. We should not conclude, therefore, that the emitting species are necessarily in the sequence of intermediates leading to the physiological reaction.

Temperature Effects

Since electronic transitions can proceed regardless of the vibrational or rotational state of the absorbing molecules, pure photochemical reactions proceed equally rapidly at all temperatures. Zero temperature dependence becomes then an identifying feature of pure photochemical reactions or the photochemical part of mixed photochemical and dark reactions. Low temperatures can of course modify absorption spectra and therefore action spectra.

Electron Spin Resonance

Metastable excited states result when an excited singlet passes to a state that can be reached from the ground state only by a forbidden transition. The most familiar cause of forbiddenness is the rule that the number of unpaired electrons must be unchanged by a transition. In most molecules (O_2 is an exception), electron spins are scrupulously paired in the ground state. We may write such a spin configuration as $A_{\uparrow\downarrow}$. Allowed transitions preserve spin pairing; forbidden transitions change it. Thus

$$A_{\uparrow\downarrow} \rightarrow A^*_{\uparrow\downarrow} \text{ or } A_{\uparrow\downarrow} \rightarrow A^*_{\uparrow\downarrow}$$

would be allowed, but

$$A_{\uparrow\downarrow} \rightarrow A^*_{\uparrow\downarrow}$$

is forbidden.

Now unpaired spins interact with magnetic fields of the appropriate strength and may be detected as perturbations of an imposed field as each electron takes a quantum jump to a paired configuration. This method known as electron spin resonance (ESR) or electron paramagnetic resonance, can be employed to detect free unpaired electrons. We can infer from an ESR signal that a sample contains free radicals or triplet states and can also infer something about the abundance, the rates of decay, and the environment of the unpaired electrons. ESR spectrometers display their signals as derivatives with the applied magnetic field strength (H_o) as the abscissa. The *g-value*, the field position at which maximum absorption occurs is related to the magnetic field strength by the relation

$$g = \frac{hv}{\beta H_0}$$

where h is Planck's constant, is the frequency of the microwave radiation (typically 9.5 kMHz), β is the Bohr magneton, and H_o is the field strength. Most absorption occurs near g=2 where H_o is about 3,400 gauss. Other identifying features of ESR spectra are the width of the signal, ΔH, measured in gauss between points of maximum slope, the amplitude, and the total signal strength, which is proportional to the number of spins.

We know that for many plants the Calvin-Benson hypothesis is approximately correct. In order for carbon dioxide to be reduced to the level of sugar in this scheme, there must be a source of ATP and NADPH. We inferred further that we must look to the light reaction of photosynthesis for a means of generating these two coenzymes. Directly or indirectly there must be reactions

$$ADP + P + hv \rightarrow ATP \quad ...(9)$$

$$NADP^+ + H + hv \rightarrow NADPH + H^+ \quad ...(10)$$

in which photon energy *hv* drives these two reaction. In most green plants the source of hydrogen in Equation 10 is water, and we may rewrite Equation 10 as follows:

$$NADP^+ + H_2O + hv \rightarrow NADPH + \tfrac{1}{2}O_2 + H^+ \quad ...(11)$$

In this way see that oxygen, a familiar product of photosynthesis, arises as a waste product. From the study of the comparative biochemistry of photosynthesis, van Niel (1949) proposed that bacteria, algae, and higher plants shared a more general process of photosynthesis, which we may write as

$$NADP^+ + H_2A + hv \rightarrow NADPH + A + H^+ \quad ...(12)$$

To include the reduction of CO_2 We must write

$$CO_2 + 2\ H_2A \rightarrow (CH_2\ O) + H_2O + 2A \quad ...(12a)$$

For bacteria and some algae, the hydrogen conor H_2A need not be water, but can be any of several substances: thiosulphate, H_2S hydrogen gas, or organic substances such as succinate. The fundamental question to be asked are the same as those for any photochemical reaction. What is the absorbing pigment? What is the fate of the pigment in the excited pigment? What are the immediate chemical products of the reaction?

In the present case we must also pursue what may be the essentially biochemical question of the formation of ATP and NADPH form the immediate products of photochemistry. It has long been recognized that photosynthesis is one of the most remarkable processes in nature. At least 30 percent of the energy of light can he transformed by photosynthesis to chemical energy an efficiency which is some 10 times greater than any other known photochemical reaction. Because of the enormous theoretical importance of the light reaction for biology, physics, and chemistry, it has attracted some of the most vigorous minds of these three disciplines. First, a word about experimental systems the two broad categories of plant material that have been commonly used are suspensions of unicellular algae and chloroplasts isolated from higher plants. Students of photosynthesis have employed algae from every algal phylum to advantage, and also photosynthetic bacteria.

Unfortunately, the algae have not yielded active chloroplasts so readily; consequently, *in vitro* studies have rested for the most part on chloroplasts from a few especially cooperative plants, such as spinach, swiss chard, and poke weed (*Phytolacca americana*). The chromatophores of bacteria, such as *Rhodospirillum rubrum*, have also been helpful, although these may be *strongly* modified during cell disruption. Because gas exchanges can be measured sensitively and conveniently, the rates of the light reaction are usually measured as CO_2 absorption and oxygen consumption. They are not necessarily equal. In some cases ATP and NADPH formation have been measured directly.

Hypothesis

The simpler photochemical hypothesis of photosynthesis was that chlorophyll is the absorbing pigment and that excited chlorophyll causes the dissociation of water to yield reducing equivalents and oxygen. Braces were placed discreetly about the H and OH to indicate our innocence of their state. The scheme was elaborated further by the addition of X and Y to serve as intermediate acceptors for H and OH. For example,

$$HOH + XY \rightarrow XH + YOH$$

The supposition of X and Y has had some heuristic value but has not of itself advanced our understanding of photosynthesis. The fundamental notion here is that of charge separation. The first photochemical act can be thought of as the separation of a negative and positive charge, as could occur in a photoionization. If the photosynthetic machinery can be thought of as a crystalline array, we can substitute the idea of electrons and holes of a semiconductor.

The proposed action of the excited pigment is analogous to that of a compressed spring. Excitation energy compresses the spring; the spring then pulls the charges apart. The negative and positive charges or electrons and boles are then stabilized as reductant (NADPH) and oxidant (O_2), or alternatively they recombine along an electron transport pathway, which is coupled to the phosphorylation of ADP. This simple picture serves as a fairly consistent model for bacterial photosynthesis. According to one hypothesis, a long wavelength form of bacteriochlorophyll termed P_{870} (short for "pigment absorbing at 870 mμ") is the photoreactive pigment. It forms a complex with an electron acceptor or oxidant, possibly coenzyme Q, and a cytochrome.

Upon excitation, P_{870} photoionizes and transfer an electron to COQ. An electron is then donated to the now-oxidized P^{+}_{870} by the accompanying cytochrome. The separated charges then recombine along the postulated electron transport chain with the coupled generation of ATP. The idea of photochemical charge separation has been found as fruitful with green plants as with bacteria, but here the hypothesis of a single photoreactive pigment supplying energy to a single electron transport system is clearly inadequate. Let us consider a model that incorporates the idea of charge separation in two photochemical systems.

Two-pigment Hypothesis

The two-pigment hypothesis is directly traceable to Robert Emerson's studies on action spectra. He inferred from his investigations that two pigments bad to be excited for photosynthesis to occur. Among the possible arrangements of a system of two pigments, Hill and Bendall (1960) proposed a "series formulation" or "z-scheme" very close to that currently accepted by most workers. In this scheme one pigment, called P_{700}, absorbs at about 700mμ, and another which we shall call P_{670}, absorbs at 670 mμ. P_{700} and an associated electron transport chain compose "system I," P_{670} and its electron transport chain compose "system I," P_{670} and the reducing of system II acts us a bridge between the two. The better part of system I is directly analogous to the whole of the bacterical charge separation system. There is an important difference, however: the electron acceptor, or oxidant, X is sufficiently electronegative to reduce NADP. The movement of electrons and "holes" or positive charges proposed for system I is

$$\text{cyto f} \leftarrow \overset{\oplus}{P}_{700} \overset{*\ e^-}{\rightarrow} X \overset{e^-}{\rightarrow} \text{ferredoxin} \overset{e^-}{\rightarrow} \text{NADP}$$

Ferredoxin is an iron-containing protein with an unusual structure.

System 11 is thought to consist of components with more positive (that is, more oxidizing) redox potentials, whose function is to oxidize OH^- to O_2. This strongly oxidizing system contains a manganese component, labeled here as E_{Mn}. The electron acceptor is called "Q" for "quencher." The electron transport system following Q probably contains plastoquinone (PQ), which then reduces cytochrome. PQ is one of a number of CoQ-like quinones fcund in green plants. Note that the role proposed for quinones is substantially different from that in bacterial photosynthesis. The electron and hole transport system of II is

$$OH^- \leftarrow \overset{\oplus}{E}_{Mn} \leftarrow \overset{\oplus}{P}_{670} \overset{*\ e^-}{\rightarrow} Q \overset{e^-}{\rightarrow} PQ \overset{e^-}{\rightarrow} \text{cyto } \overset{e^-}{b} \rightarrow \text{cyto f}$$

Phosphorylation may in principle be coupled to one or both of the electron transport systems, but most workers place it on the electron transport "bridge" connecting Q and cytochrome *f.* The systems of the two-pigment hypothesis can be conceived of as photochemical specialization.

The generation of reduced NADP and the utilization of water as a source of electrons presented two separate evolutionary problems. Once Nature had mastered the trick of photochemical charge separation in purple bacteria, she evolved distinct but interacting photochemical systems toward the solution of these separate problems. The result was green plants with a photosynthetic versatility far surpassing that of bacteria. Bearing in mind that these schemes, in addition to being complete, may also be wrong, let us turn to the evidence.

Bunsen-Roscoe Law

Obedience to the Bunsen. Roscoe law is an essential characteristic of true (or at least uncomplicated) photochemical reactions. If we are to study the photochemistry of processes in which a photoreaction is accompanied by "dark" chemistry, it is essential to limit ourselves to experimental conditions under which the dark reactions are not limiting to the overall process. Plants typically show a fairly wide range of light intensities over which the rate of photosynthesis (measured as CO_2 reduction) in linear. Beyond some critical level of light which depends on the concentration of carbon dioxide, temperature, nutrition, etc., photosynthesis is no longer increased.

In 1905 Blackman proposed that the fact of interaction between light intensity and CO_2 levels was evidence for dark or enzymatic reactions intervening in photosynthesis. More conclusive evidence for dark reactions was provided by Otto Warburg, who found that the yield from high intensities of light was considerably enhanced by giving the light in brief flashes interspersed with relatively long dark intervals.

Robert Emerson and William Arnold (1932) pursued this and found that the photochemical yield was constant when dark intervals of 20 msec or longer intervened between light flashes of 0.1 msec. One might have imagined that a plant would not become saturated with light until all of its chlorophyll had become excited. But surprisingly the *number of quanta* in a saturating flash was found to correspond to only one three-hundredth of the number of chlorophyll molecules. Yet in studies with *low* light intensities, photosynthesis behaved as if all the chlorophyll were at least potentially active.

Emerson and Arnold proposed that this paradox could be resolved if a single quantum of light were "collected" in units of about 300 chlorophyll molecules. Such a "photosynthetic unit" would then process its collected light energy with a common set of enzymes. For our present purposes we may conclude that dark reactions do intervene and that the Bunsen-Roscoe law is obeyed under conditions of brief light flashes followed by relatively long dark periods.

Action Spectra

With a *single* beam of light, the effectiveness or light (dotted line) shows two principal peaks, one near 435 mµ and one near 675 mµ; the action then falls sharply to zero near 700 mµ. The action spectra obtained with a single beam, correspond only in a general way to the absorption spectrum of chlorophyll. Action is higher in green light (500 to 600 mµ,) than we should have expected from chlorophyll absorption and light beyond 700 mµ, is much less effective than the corresponding absorption of chlorophyll *in vivo.* The data are clearly at variance with the single chlorophyll pigment hypothesis first considered. The two-pigment hypothesis has its origin in Emerson's, studies of action spectra obtained with two wavelengths. Two features are observed: (a) light in the far red which cannot support photosynthesis alone *enhances* the photosynthesis when a shorter wavelength is added and (b) photosynthesis with two wavelengths together is typically more efficient than the sum of the two separately.

A sharper contrast between action spectra obtained with one and two beams of light is found with many red, blue-green, and brown algae. The action spectrum with 546-mµ supplementary light looks very similar to the absorption spectrum of chl a, but that with supplementary light of 536 mµ shows a peak in the green and minimal activity in the blue and red where chlorophyll absorbs. The inference from this phenomenon of "enhancement" or the "second Emerson effect" is that at least two photochemical reactions must occur in photosynthesis; that is. P_{700} and P_{670} must be excited in order for photosynthesis to be completed. An alternative view, that the second beam of light is required for a physical process such as excitation to a higher state, is vitiated by the finding of Blinks (1960) that the two beams of light may be separated in time. Let us look at this crucial piece of evidence more closely.

Blinks discovered that, if two beams or light at two wavelengths are given successively rather than simultaneously, 'transient changes in the rate of O_2 production, called *chromatic transients*, are observed. If darkness intervenes between the two beams, the chromatic transients still occur, but the magnitudes decrease with the length of the dark interval. The changes observed upon illumination at

680 mil show a half life of 5 sec in the dark. The lifetimes of these transients can be interpreted to mean that steady states of the pigment systems are reached within a relatively few seconds and that the intermediates which link the two systems have a lifetime of about the same duration. The action spectra for enhancement of photosynthesis obtained with two simultaneous light beams agree nicely with the action spectra for exciting chromatic transients.

One can say, then, that enhancement occurs when the two beams of light are given either simultaneously or within a few seconds of each other. There are two additional bits of intersecting evidence. A number of algae show "photoreduction," in which H_2 serves as $[H_2A]$ in Equation 12, and in which no O_2 is evolved. The action spectrum for photoreduction corresponds to the P_{700} system and does not show enhancement. From this it was, inferred that P_{670} is concerned principally with driving 0_2 evolution. Enhancement is also absent from bacterial photosynthesis, where again O_2. evolution does not occur.

Absorption of Light by Molecules

In our discussion of light utilization we will first consider the absorption process itself. Only light that is actually absorbed can be effective in producing a chemical change, a principle embodied in the Grotthus-Draper law of photochemistry. This is true whether radiant energy is to be converted to some other form and then stored or simply used as a trigger. Another important principle of photochemistry is the Stark-Einstein law, which specifies that each absorbed photon activates but one molecule. Einstein further postulated that all the energy of the light quantum is transferred to a single electron during, the absorption event, resulting in the- movement of this electron to a higher energy state. To understand light absorption, we must therefore first consider some of the properties of electrons.

Role of electrons in absorption event

From a classical view- point, an electron is a charged particle that moves in some orbit around an atomic nucleus. We can readily appreciate that the energy of the electron would depend both on the location of the orbit in space and on how fast the electron moves in its orbit. Thus, the increase in energy of an electron upon absorbing a photon could be used either to transfer that electron into an orbit which lies at a higher energy than the original orbit or to cause the electron to move more rapidly about the nucleus than it did before the light arrived, The locations of various possible electronic orbits and the speeds for electrons moving in them are both limited to certain discrete or "allowed" values, a phenomenon which has been interpreted by quantum mechanics. This means that the energy of an electron in an atom or molecule can change only by certain specific values.

Consequently, the total electronic energy of an atom or molecule occurs in discrete amounts. Only light of certain wavelengths will have the proper quantum energy to cause the electron to move from one possible energetic state to another. During light absorption an interaction occurs between the electromagnetic field of the light and some electron. Because electrons are charged particles, they will experience a force when they are in an electric field. But light consists of periodic variations in both the local electric and magnetic fields in the medium through which it is passing.

The oscillating electric field or light thus represents a periodic driving force acting on the electrons in matter. This electric field-a vector quantity which has a certain direction in space, such as along the y-axis in causes or induces electrons to move. If the frequency of the electromagnetic radiation in such as to cause a large sympathetic oscillation or beating of some electron, that electron is said to be in *resonance* with the light wave. Such a resonating electron corresponds to a so-called *electric dipole* in the molecule, as the electron is forced to move first in a certain direction and then in the opposite one in response to the oscillating electric field of the light. (An electric dipole is simply the local separation of positive and negative charge.)

The displacement of the electron back and forth requires energy–in fact, it may take the entire energy of the photon, in which case the quantum is captured or absorbed. The direction and magnitude of the induced electric dipole will depend on the resisting, or restoring forces on the electron provided by the rest of the molecule. These restoring forces depend on the other electrons and the atomic nuclei in the molecule; consequently, they are not the same in different types of molecules. The actual electric dipoles which can be induced in a particular molecule are characteristic of that molecule, which helps explain why each molecular species has its own unique absorption spectrum. The probability for the absorption of light depends on both its wavelength and the relative orientation of its electromagnetic field with respect to the possible induced oscillations of the electrons in the molecules.

Absorption of photons without ejection of electrons from the absorbing species can take place only if the following two conditions exist: (I) the light has the proper energy to get to a discrete excited state of the molecule, i.e., a specific wavelength, and (2) the electric field vector associated with the photon has a component parallel to the direction of some potential electric dipole in the molecule so that an electron can be induced to oscillate.

In other words, the electric field of light must be in the correct direction so that it is able to exert a force on some electron. It must exert this force in the direction of a potential electric dipole so that the electron can be induced to move and thus be able to accept the energy of the photon. The probability for absorption is actually proportional to the square of the cosine of the angle between the electric field vector' of light and the direction of the induced electric dipole in the molecule. Within these limits set by the wavelength and the orientation, light energy can be captured by the molecule, placing it in an excited state.

Electron spin and state multiplicity

Light absorption is affected by the arrangement of electrons in an atom or molecule, which depends among other things on a property of the individual electrons known as their "spin," We can consider that each electron is a charged particle spinning about an axis in much the same way as the earth spins about its axis. Such rotation has an angular momentum or *spin* associated with it. The magnitude of the spin of all electrons is the same; but since spin is a vector quantity, it can have different directions in space. For an electron, only two orientations are actually found, viz., the spin of the electron can be aligned either parallel or antiparallel to the local magnetic field.

Even in the absence of an externally applied magnetic field, such as that of the earth or some electromagnet, there is a local internal magnetic field provided by both the moving charges in the nucleus and the motion of the electrons. There is therefore always a magnetic field with which the electron spin can be aligned. It has proved convenient to express the spin of electrons in units of $h/2\pi$, where h is Planck's constant. In these units, the projection along the magnetic field of the spin for a single electron is either

+½ (e.g., when it is parallel to the local magnetic field), or −½ (when the spin is in the opposite direction).

The total spin of an atom or molecule is the vector sum of the spins of all the electrons, each individual electron having a spin of either + ½ or −½. The magnitude of this total spin is given the symbol S, an extremely important quantity in spectroscopy. In order to talk about the spectroscopic properties of various molecules, we will find it convenient to introduce a new term known as the *spin multiplicity*. The spin multiplicity of an electronic state is defined as 2S + 1, where S is the magnitude of the net spin for the whole atom or molecule. For example, if S equals zero–which means that the spin projections of all be electrons taken along the magnetic field cancel each other–then 2S + 1 is one, and the state is called a *singlet*. On the other hand, if S equals one, the state is a *triplet* (2S+ 1 then equals 3). Singlets and triplets are the two most important spin multiplicities encountered in biological systems, and we will consider both in further detail below.

When referring to an absorbing species, the spin multiplicity is generally indicated by S for singlet and T for triplet, as we will do here, When S equals ½ as would occur if there were one unpaired electron in a molecule, 2S + 1 is 2; such "doublets" occur for free radicals.) Electrons are found of course only in certain "allowed" regions of space; the particular locus in which some electron can move is referred to as its *orbital*. In the 1920s W. Pauli observed that when an electron is in a given atomic orbital, a second electron having its spin in the same direction is excluded from that orbital. This led to the enunciation of the *Pauli exclusion principle* of quantum mechanics: when two electrons are in the same orbital, their spins must be in opposite directions. When a molecule has all of its electrons paired in orbitals with their spins in opposite directions, the total spin of molecule is zero (S = 0), and the molecule is in a singlet state (2S + 1 = 1). The ground, or unexcited state of most molecules is such a singlet, i.e., all the electrons are in pairs in the lower energy orbitals. When some electron is excited to an unoccupied orbital, two spin possibilities exist.

The spins of the two electrons (which are now in different orbitals) may be in opposite directions, as they were when paired in the ground state. This electronic configuration is still a singlet state. On the other hand, the two electrons may have their spins in the same direction, a triplet state. (Since the electrons are in different orbitals, their spins can be in the same direction without violating the Pauli principle). An important rule–first enunciated by Hund, based on empirical observations, and later explained using quantum mechanics–is that the level with the greatest spin multiplicity has the lowest energy. Thus, an excited triplet state is lower in energy than the corresponding excited singlet state, we will consider this phenomenon for chlorophyll.

Molecular orbitals

For a discussion of the light absorption event involving the interaction of an electromagnetic wave with some electron, it is easiest to visualize the electron as a small point located at some specific position in the atom or molecule. For a classical description of the energy of an electron, we can imagine the electron moving in some fixed trajectory or orbit about the nucleus. However, to describe the role of electrons in binding together atoms to form a molecule, it is most convenient to imagine the electron spatially distributed like a cloud of negative charge surrounding the nuclei of adjacent atoms in the molecule. In this last description, involving probability considerations introduced by quantum mechanics, we say that the electrons are located in *molecular orbitals.* In describing the electronic states of molecules we need to use the language of molecular orbitals in order to discuss adequately the absorption of light by chlorophylls, carotenoids, and other pigments in plants.

We must therefore introduce many new terms and unfortunately, many new concepts which can truly be mastered only by a study of quantum mechanics. Our common experience yields relations such as Newton's laws of motion, which describe events on a scale much larger to atomic or molecular. By contrast, the main application of quantum mechanics is to molecular, atomic, and subatomic dimensions. On such a scale our intuition often fails us. Moreover, may things which otherwise seem to be absolute, such as the position of an object, are really describable only as a probability. We cannot say that an electron is located at such and such a place.

Instead, we must be satisfied with knowing only the *probability* of finding an electron in the region of space about some nucleus. A logical way to begin our discussion of molecular orbitals is to consider the probabilities of finding electrons in given regions about a single atomic nucleus. The region of space in which an electron moves is of course its orbital. The simplest case is the hydrogen atom which contains a single electron moving about a nucleus. If we were to determine the probability of finding that electron in the various regions of space about its nucleus, we would find that it spends most of its time fairly close (within a few Angstroms) of the nucleus, and that the probability distribution is *spherically symmetric* in space. In other words, the chance of finding the electron is the same in all directions about the nucleus.

In atomic theory, this spherically symmetric distribution of the electron about the hydrogen nucleus is called an *s* orbital, and the electron is referred to as an *s* electron. The next simplest case is the helium atom. Helium has two s electrons moving in the *s* orbital about its nucleus. By the Pauli exclusion principle these two electrons must have their pins in opposite directions, as is indeed the case. What happens for lithium, which contains three electrons? Two of its electrons are in the same type orbital as that of helium. This orbital is known as the K shell, and represents the spherically symmetrical orbital closest to the nucleus. The third electron is excluded from the K shell but occurs in another orbital with spherical symmetry whose probability description indicates that on the average its electrons are further away from the nucleus than is the case for the K shell. Electrons in this new orbital are still *s* electrons, but they are now referred to as being in the L shell. Beryllium, which has an atomic number of four, can have two *s* electrons in its K shell and two more in its L shell.

What happens when we have 5 electrons moving about a nucleus, as in an uncharged boron atom? It turns out that the fifth electron is in an orbital that is not spherically symmetric about the nucleus. Instead, the probability distribution is in the shape of a dumbbell, although it is still centered about the nucleus. This new orbital is referred to as a *p* orbital and electrons in it are called *p* electrons. The probability of finding a *p* electron in various regions of space is greatest along some axis through the nucleus and vanishingly small at the nucleus itself. How do all these various atomic orbitals relate to the spatial distribution of electrons in molecules?

A molecule contains more than one atom (except for "molecules" like helium or neon) and certain of the electrons can move between the atoms. Such interatomic motion is of course crucial for holding the molecule together. Fortunately, it still proves possible to describe the spatial localization of electrons in molecules by using suitable linear combinations of the spatial distributions represented by various atomic orbitals centered about the nuclei involved. In fact, molecular orbital theory is concerned with giving the correct quantum-, or wave-mechanical description of the probability of finding electrons in various regions of space in *molecules* by using the already known descriptions for the probability of finding electrons in various locations making up their *atomic* orbitals. Some of the electrons in molecules are localized about a single nucleus, while others are *delocalized*, or shared between nuclei.

For the delocalized electrons, the combination of the various atomic orbitals used to describe the spatial positions of the electrons in three dimensions is consistent with a sharing of the electrons between adjacent nuclei. In other words, the molecular orbitals of the delocalized electrons spatially overlap more than one nucleus. This sharing of electrons is responsible for the chemical bonds that prevent the molecule from separating into its constituents atoms, i e., the negative electrons moving between the positive nuclei told the molecule together by attracting the nuclei of different atoms. Moreover, these delocalized electrons are the ones usually involved in light absorption by molecules. The lowest energy molecular orbital is σ orbital, which is analogous in symmetry properties to the *s* atomic orbital.

Nonbonding, or lone pair electrons contributed by atoms like oxygen or nitrogen occur in *n* orbitals and retain their atomic character in the molecule. These *n* electrons are essentially physically separate from the other electrons in the molecule and do not take part in the bonding between nuclei. We shall devote most of our attention to electrons in π molecular orbitals, which are the molecular equivalent of π electrons in atoms. Such π electrons are delocalized in a bond joining two or more atoms, and they are of prime importance in light absorption. In fact most photochemical reaction and spectroscopic properties of biological importance result from the absorption of photons by the π electrons. The excitation of a π electron by light absorption can lead to an excited state of the molecule in which the electron moves into a π* orbital, the asterisk referring to an excited, or high energy molecular orbital. The probability distributions for electrons in both π and π* orbitals (the circumscribed regions indicate where the electrons are most likely to be found), a π orbital delocalized between two nuclei: the same clouds of negative charge electrostatically attract both nuclei. Such sharing of electrons in the π orbital helps join the atoms together, hence we refer to this type of molecular orbital as *bonding*.

Electrons in a π^* orbital do not help join atoms together, rather, they tend to decrease bonding between atoms in the molecule, since the clouds of negative charge in adjacent atoms are located in a position to repeal each other. A π^* orbital is therefore referred to as *antibonding.* The decrease in bonding in going from a π to a π^* orbital results in a less stable (higher energy) electronic state for the molecule, so that a π^* orbital is at a higher energy than a π orbital. In photochemistry the energy required to move an electron from the attractive (bonding) π orbital to the antibonding π^* orbital is obtained by the absorption of a photon. For molecules such as chlorophylls and carotenoids, which we will consider later, the π^* orbitals are often only a few electron volts higher in energy than the corresponding π orbitals.

Light energy can cause molecular changes known as *photo-isomerizations.* The three main types of photoisomerization are (I) *cis-trans* isomerization about a double bond, (2) a double bond shift, like that implicated in the interconversion between the two forms of phytochrome that we will present below, and (3) a molecular rearrangement involving changes in carbon-carbon bonds, *e.g.*, ring cleavage or formation. *Cis-trans* photoisomerization–which illustrates the consequences of the different spatial distribution of electrons in π and π^* orbitals–makes possible the generally restricted rotation about a double bond.* A double bond has two π electrons; light absorption leads to the excitation or one of them to a π^* orbital. The attraction between the two carbon atoms caused by the remaining π electron is mostly canceled by the antibonding, or repulsive contribution from this π^* electron.

Hence, the molecular orbitals of the original two π electron, which prevented rotation about the double bond have been replaced by an electronic configuration permitting relatively easy rotation about the carbon-carbon axis. Let us consider the absorption of light by a molecule in the *cis* form. When the excited π^* electron drops back to a π orbital, the two large groups can be on the same side (the original cis form) or on opposite sides (the *trans* isomer) of the double bond. Since the two isomers can have markedly different properties, such use of light to trigger their interconversion can have profound importance in biology. For example, light can cause the photoisomerization of a *cis* form of retinal yielding a *trans* isomer, which is the basic photochemical event triggering vision.

Light absorption by chlorophyll

To illustrate certain of the terms introduced above to describe the absorption of light by molecules we will use chlorophyll. The principal energy level and electronic transitions of chlorophyll are presented. Chlorophyll is a singlet in the ground state, as are practically all pigments of importance in biology. Upon the absorption of a photon, some π electron is excited to a π^* orbital. When this excited state is a singlet, it is represented by $S_{(\pi. \pi^*)}$. The first symbol in the subscript is the type of electron (here a π electron) which has been excited to the antibonding orbital indicated by the second symbol (here π^*). We will represent the ground state by $S_{(\pi, \pi)}$ indicating that no electrons are in excited, or antibonding orbitals.

If the orientation of the spin of the excited π electron has become reversed during excitation, it would be in the same direction as the spin of the electron which remained in the π orbital. (Each filled orbital contains two electrons whose spins are in opposite directions.) In this case, the net spin of the molecule in the excited state would be 1, known as a triplet. The excited triplet state of chlorophyll is represented by $T_{(\pi, \pi^*)}$. Chlorophyll has two principal excited singlet states which differ considerably in energy. One of these states, designated $S^a{}_{(\pi, \pi^*)}$ can be excited by red light, e.g., a wave length of 680 nm. The other state, $S^b{}_{(\pi, \pi^*)}$ lies higher in energy and is excited by blue light (e.g. 430 nm).

The electronic transitions caused by the absorption of photons arc indicated by solid vertical arrows, while the vertical distances correspond approximately to the differences in energy involved. Excitation of a singlet ground state to an excited triplet state is usually only about 10^{-5} times as probable as going to an excited singlet state, so that the transition from $S_{(\pi, \pi)}$ to $T_{(\pi, \pi*)}$ has not been indicated for chlorophyll. In order to go from $S_{(\pi, \pi)}$ to $T_{(\pi, \pi^*)}$, the energy of the electron must be substantially increased and the orientation of the electron spin must be simultaneously reversed. Since the coincidence of these two events is rather improbable, very few chlorophyll molecules are directly excited to $T_{(\pi, \pi*)}$ from the ground state by the absorption of light. We will now briefly compare the two most important excitations in photobiology, *viz.,* the transitions of *n* and electrons to π^* orbitals. The *n* electrons have very little spatial overlap with other electrons, and they tend to be higher in energy than the π electrons, which are reduced in energy (stabilized) by being delocalized over a number of nuclei.

Therefore, the excitation of an *n* electron to a π^* orbital generally takes less energy than the transition of a π electron to the same antibonding orbital. Hence, a $S_{(n, \pi^*)}$ state usually occurs at a lower energy–a longer wavelength is required for its excitation–than does the analogous $S_{(\pi, \pi^*)}$. Another difference is that π and π^* orbitals may overlap spatially, while n and π^* orbitals generally do not. Consequently, the *n* to π^* transition is not as favoured or probable as the π to π^* one. Thus, transitions to $S_{(\pi, \pi^*)}$ states tend to dominate the absorption properties of a molecular species (as evidenced, for example, by an absorption spectrum) compared with excitations yielding $S_{(n, \pi^*)}$ states. As our final topic in this section, we will consider the effect of molecular environment on n, π and π^* orbitals.

The *n* electrons generally interact strongly with water, e.g., by participating in hydrogen bonding, and therefore the energy level of *n* electrons is considerably lower in an aqueous environment than in an organic solvent. The energy of π^* electrons is also lowered by water, but to a lesser extent than for *n* electrons; the π orbitals ate the least affected by the solvent. Thus, a transition from an *n* to a π^* orbital takes more energy in water than in an organic solvent. On the other hand, the π to π^* transition takes less energy in an aqueous environment than in an organic one. As an example, let us consider the transition of chlorophyll to an excited singlet state in different solvents. The excitation to either excited singlet state takes less energy in an aqueous environment than it does in an organic solvent, i.e., when chlorophyll is in water the required photons have a longer wavelength than when the chlorophyll is dissolved is an organic solvent like acetone. This occurrence of a lower energy requirement in water is consistent with our statement that the transition for chlorophyll corresponds to the promotion of a π electron to a π^* orbital.

Light Emission

If the photosynthetic pigments can be persuaded to emit some of their excitation energy, we can hope to identify them both from their fluorescence emission spectra and action (that is, fluorescence excitation) spectra. In fact plants can be seen to fluoresce, especially when photosynthesis is inhibited or prevented, just as if the excitation energy were switched from photochemistry to light emission. The emission spectrum of green leaves and algae at room temperature normally shows a single maximum at 685 mμ.

At liquid nitrogen temperature the emission band splits into two or three bands at 685, 695, and 720 mμ. The location and widths of the absorption bands at 685 and 695 mμ are characteristic of chlorophyll *a* in organic solvents, In fact fluorescence of chl *b*, carotenoids, or phycobilins is rarely observed in vivo. There are, however, long-wavelength fluorescence bands which might be due to long-wavelength forms of chl *a* in P_{700}. In sharp contrast to the unanimity of the emission spectra of plants, the excitation spectra are far more diverse. Chl *a* fluorescence can be induced by exciting chl *b*, carotenoids phycobilins, and chl *a* itself. If these other pigments are excited and chl *a* emits the fluorescence, we can infer only that there must be energy transfer (exciton migration) among them and that chl *a*, or at least some form of it, is the energy sink.

Quantitatively, the excitation spectra show that most fluorescence comes from system II. Indeed, the oxidant of system II was named "quencher" (Q) because in its normal, oxidized state it competes with or quenches fluorescence for the energy of P_{670} Consistent with this view, one finds that fluorescence is decreased by absorption of light by system I, as this causes QH to revert rapidly to Q, and is promoted by DCMU which blocks the oxidation of QH. As suggested by Duysens and Amesz (1962) the model can be refined further: the two pigment systems are each arrays of pigments with overlapping dipoles. System I is composed of about 300 (chl *a* 683): I (cyt f): I (chl *a* 700) at the energy sink, or collection center. System II is composed of chl *b* or, in the case of blue-greens, red, and brown algae, phycobilins be energy sink in this case is another form of chlorophyll, chl a 673, at the collection center. Hundreds of pigment molecules light things be combined into two separate kinds of units with overlapping dipoles [as in the photosynthetic units of Emerson and Arnold (1932)], so that a photon absorbed by anyone of the molecules in the unit is passed on to a few molecules of chlorophyll *a* at the collection center

The average distance between chlorophyll molecules in chloroplasts is computed to be 15 to 20Å. At this distance the model for energy transfer predicts nearly 100 percent efficiency. Additional light-gathering or "antenna" pigments include β-carotene, which probably transfers only to system I. Xanthophylls appear not to transfer at all. What if one collection center becomes inoperative, due, for example, to a bleached reaction center? Can photons be passed from one unit to another of the same kind? The answer appears to be yes. On the basis of fluorescence measurement units of system I or system II can pass photons to other units with an efficiency of a about 0.5. A more difficult problem, both conceptually and experimentally, is that called "spill-over," the transfer of energy between different kinds of units.

If system II were to receive more light energy than it could efficiently use, could it relay the excess energy to system I? The seeming necessity for some regulated spill-over comes from the nearly constant quantum yield of photosynthesis over a range of wavelengths where the two systems would be expected to absorb very unevenly. The experimental evidence points both ways. Experiments with chromatophores of *Chromatium*, one of the photosynthetic bacteria, support this general scheme. These bacteria ha ye at least two kinds of bacteriochlorophyll, B_{896} and B_{856}. When the chromatophores have been extracted with detergents, fluorescence of B_{850} is enhanced whereas that of B_{890} is decreased, as if the detergent had interrupted exciton migration to the long-wavelength energy sink. Blue fluorescence characteristic of NADPH or NADH emission also occurs. The action spectra for NADPH fluorescence corresponds to that of P_{700}.

Electron Spin Resonance

Electron spin resonance can be used to detect unpaired electrons, such as those that might exist in free radicals or triplet states. Signals from photosynthesizing cells were first detected by Commoner's group. Two finds of ESR signals (originally denoted I and II) are seen in green plants. Signal I has a g-value of 2.0025 and a ΔH of 7 to 9 gauss. A closely similar signal occurs in photosynthetic bacteria. The occurrence of signal I correlates exactly with P_{700} (or the corresponding bacterial reaction centers) with respect to the kineties of their response to light their response to redox potential and their deletion from *Scenedesmus* mutants and at least approximately with the estimated numbers of reaction centers. Although chlorophyll alone will give a similar signal *in vitro*, there is no correlation *in vivo* with bulk chlorophyll. Signal II just as fortuitously correlates with system II. It has a g-value of 2.0046, and its ΔH value is about 20 gauss. It is induced by *light*, as is signal I, but decays over several hours in the dark.

Signal II is absent from bacteria. From the absence of signal II from *Screnedesmus* mutants lacking plastoquinone and the similarity of the ESR signal of plastoquinone *in vitro*, it is not unreasonable to think that a semiquinone of plastoquinone may be the source of signal II.

Conclusions on Photochemistry

The accumulated evidence is consistent with the overall view of the two-pigment hypothesis. The participation of a variety of chloroplast pigments as collectors of photon energy is clearly indicated, but special forms of chl *a*, probably bound to specific proteins appear to be the photoreactive components. Thus the identities of the absorbing pigments arc tentatively, and approximately established. Photoionization of the excited chl *a* centers, followed by stabilization of the charge separation, is established with about the same level of uncertainty. It seems clear from the lack of temperature dependence of the absorption changes corresponding to cytochrome oxidation that electron transport is set in motion as an early event after excitation. However, we must reserve judgment on the precise nature of the photoreaction.

The behaviour of system I, including P_{700}, is reasonably clear for the simple reason that it can be set in motion by light which leaves system II cold, so to speak. The most promising approaches to the isolation of system II are through excitation of the phycobilins of red algae where the overlap between systems I and II is minimal, and through the use of algal mutants, in which the interaction has been, rendered inoperative. Until complete stories are filed on several organisms, we must view our formulations with cheerful skepticism.

Biochemical Evidence

We should now assemble the biochemical evidence for the pathways of ATP and NADPH formation. Such evidence has been obtained largely with isolated chloroplasts and fragments of chloroplast lamellae, loosely called grana, from higher plants. Additional evidence has come from the method of mutations. Robert Hill (1939) first demonstrated that oxygen could be liberated from water by illuminated chloroplast provided with a variety of hydrogen acceptors.

$$B + H_2 \xrightarrow[\text{chloroplasts}]{hv} BH_2 + \tfrac{1}{2}O_2$$

where B could be ferricyanide, benzoquinone, or any of a variety of dyes lodging by the action spectra of the Hill reaction, ferricyanide must normally accept electrons from system II. It thus provides a lever for separating the O_2 liberating end of photosynthesis from the photochemistry of system I. But the Hill reaction is at once more complicated and more general than system II. It now seems that Hill oxidants, depending on their redox potentials, may act as electron acceptors either to X^- or to the P_{670} electron acceptor or to other intermediates in electron transport.

Hill's general strategy of feeding oxidants to photoreducing systems helped others to identify the electron transport from X to NADP through the isolation and characterization has a most unusual structure for a protein. These steps have been worked out by San Pietro, Jagendorf, Arnon, and others. Formation of the other required photoreaction product, ATP can also be demonstrated in isolated chloroplasts and bacterial chromatophores. The process, called ***photophosphorylation***, was first described by Arnon *et. al.* (1954) with chloroplasts and by Frenkel (1954) with fragments ("chromatophores") of photosynthetic bacteria. Arnon identified both "cyclic" and "noncyclic" photo-phosphorylation.

In the cyclic process, phosphorylation is somewhat coupled to an electron transport chain that starts with one of the photo acts (probably P_{700}); the electrons proceed around a closed path, returning to chlorophyll. In the noncyclic process, the electron chain terminates in a pool of oxidant, *e.g.,* NADP. Which of these two processes fuels the Calvin-Benson cycle of CO_2 fixation? Tanner *et. al.* (1969) have approached this question by showing that noncyclic (but not cyclic) photophosphorylation proceeds ***pari passu*** with CO_2 fixation with respect to light intensity and the response to certain kinds of inhibitors.

While cyclic phosphorylation surely exists *in vivo* and indeed seems to be the sole contributor of ATP to several transport processes, its magnitude appears to be too small to contribute significantly to normal rates of photosynthesis. The difficulty, as in the case of absorption changes, is that different workers obtain substantially different results depending on the conditions and the choice of experimental material. Action spectra for photophosphorylation obtained in different laboratories show peaks at the following wavelengths: 650, 670, 677, 680, and 710 mμ. In other words, the action spectra correspond to system I, system II, and both! The simplest conclusion for the present is that photophosphorylation may be coupled to any of the electron transport chains in the chloroplast.

2

PROCESS OF PHOTOSYNTHESIS

The process in which certain carbohydrates are synthesized from carbon dioxide and water by chlorophyllous cells in the presence of light, oxygen being a by-product, is generally called ***photosynthesis.*** On the continent of Europe and to a lesser extent in Great Britain,the terms ***carbon assimilation*** and ***assimilation*** are often used to designate this process. The common use of the word "assimilation" to denote the process in which foods are incorporated into the structures of the plant body makes the employment of this term or carbon assimilation as a synonym for photosynthesis undesirable.

The *summary* equation representing the process of photosynthesis is conventionally, written as follows:

$$6CO_2 + 6H_2O \xrightarrow[\text{chlorophyllous cells}]{\text{673 kg.-cal of radiant energy}} C_6H_{12}O_6 + 6O_2$$

It should be clearly understood that this equation does not support to indicate the mechanism of the process, but is merely a statement of the net exchanges which occur in the shorthand of chemical symbols. As usually written, a hexose sugar is designated as the carbohydrate formed in photosynthesis, but this is by no means certain. The value 673 kg.cal. is, strictly speaking, correct only for glucose; the energy requirement of the reaction is approximately the same, per CH_2O group, however, for the synthesis of any of the hexoses,or for any disaccharide or polysaccharide derived from hexoses. Photosynthesis (Gr. *Photos* light; *synthesis* = putting together) is an anabolic process of manufacture of organic compounds inside the cell, containing chlorophyll, from carbon dioxide and water with the help of sunlight as an energy source. Basically, photosynthesis is the absorption of light energy and its conversion into chemical potential.

Historical Background

1. *Aristotle and Theophrastus* (320 B.C). Thought that plants derive all their nourishment directly from soil.
2. *Hales* (1727). Concluded that plants are nourished, in part, from atmosphere.
3. *Joseph Pristley* (1772). Discovered the evolution of oxygen by plants. He also showed that oxygen is consumed by animals.
4. *Janingen-Housz* (1779-96). Purification of air is carried out by green plants only and that light is essential to the process. He associated these reactions with the green parts of plants.
5. *Jean Senebier* (1782). The rate of purification of air by green plants depends upon the concentration of foul air.
6. *Nicolas de Saussure* (1804). Noted that water is involved in photosynthesis, plants pick up carbon dioxide and release oxygen.
7. *Dutrochet* (1837). Established connection between photosynthesis and chlorophyll.
8. *Leibig* (1845). Organic matter synthesized during photosynthesis is derived from dioxide and water.
9. *Von Mayer* (1845). Green plants convert solar energy into chemical energy during photosynthesis.
10. *Sachs* (1862). Proved that starch is synthesized by plants in light dependent reaction i.e., photosynthesis.
11. *Englemann* (1882). Showed that red light is the most efficient stimulator of photosynthesis.
12. *Meyer* (1883). First described details of chloroplast structure.
13. *Blackman* (1905). Propounded the principle of limiting factors. He also proposed the occurrence of a dark phase in photosynthesis.
14. *Wilstaller and Stoll* (1913). Isolated chlorophyll and later determine its structure.
15. *Warburg* (1919). Found that the efficiency of photosynthesis is higher with intermittent light.
16. *Warburg and Negelein* (1922). First measured the efficiency of photosynthesis.

17. *Thunberg* (1923). Recognized that carbon dioxide is reduced and water oxidized during photosynthesis.
18. *Emerson and Arnold* (1932). Performed flashing light experiments at different temperature and found out the existence of light and dark phases of photosynthesis.
19. *Wood and Werkman* (1936). Distinguished between the light reaction and dark reaction.
20. *Hill* (1938). Evolution of oxygen occurs in light reaction performed by isolated illuminated chloroplasts in the presence of suitable electron acceptors and the absence of carbondioxide.
21. *Ruben, Randall, Kamen and Hyde* (1941). Showed that the oxygen liberated during photosynthesis comes from water and not from carbon dioxide.
22. *Calvin and Benson* (1948). Showed that phosphoglycerate is an early product of carbon dioxide fixation.
23. *Calvin* (1954-55). Traced the pathway of carbon dioxide fixation by using $^{14}CO_2$ and gave the C_3 cycle,now known as *Calvin Cycle. Calvin* was awarded Nobel Prize in 1960.
24. *Hatch and Slack* (1965). Discovered supplementary mechanism of carbon dioxide fixation called C_4 pathway in certain topical plants.
25. *Huber et al* (1985). Crystallized and analysed the photocentre of *Rhodobactor* through X-ray differection technique. They were awarded Nobel Prize in 1988.

The Structure of the Leaf

Since the leaf is the chief site of the photosynthetic process, information on the structure of this organ is fundamental to a full understanding of the leaf activities. The leaf is commonly studied in cross-section–that is by cutting through the leaf at right angles to the midvein. The tissues of the leaf may be classifed as (1) *epidermis,* (2) *mesophyll* and (3) *vascular bundles.*

Epidermis

The epidermis is composed of a single layer of interlocked cells and contains no chloroplasts. It is continuous over both surfaces of the leaf and is therefore distinguished into upper and lower epidermis. It is covered by the *cuticle,* a waxy layer, or film, which retards the movement of water and gases into and out of the leaf. The cuticle varies in thickness in different species; when it is thick, it is not appreciably penetrated by oxygen and carbon

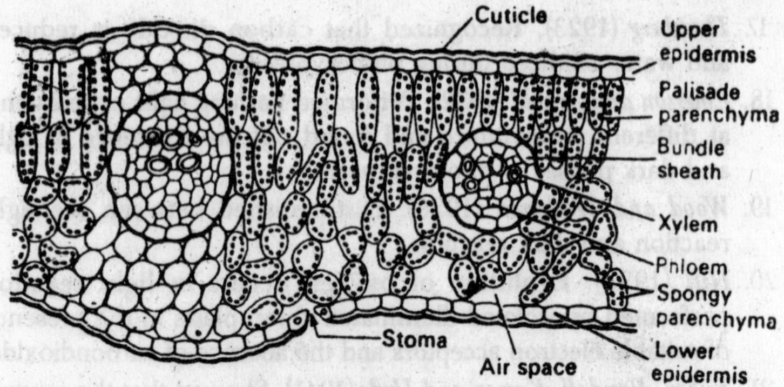

Fig. 2.1. Detailed structure of a part of T.S. of mango leaf.

dioxide, but in leaves with thin cuticles there is some evidence of movement of gases through the epidermal cells. Similarly, although the cuticle impedes the loss of water vapour from the internal tissues, such loss, under certain conditions, may nevertheless be considerable. The epidermis is perforated by numerous lens-shaped openings, called *stomates*, which open into large intercellular spaces within the leaf. Stomates are also found in the epidermis of young twigs and herbaceous stems. The stomate is really an intercellular space formed by the splitting of the cell wall between two specialized epidermal cells known as *guard cells*. These, unlike other epidermal cells, contain chloroplasts. The guard cells, in surface view, are typically crescent-shaped.

In cross-section, the inner walls of the *guard cells*, those which face the stomate, are seen to be much thickened. The upper and lower corners of the walls adjacent to the stomate are also thickened and frequently form projecting ridges. As a result of these thickenings and the shape of the guard cells, changes in turgor within them operate to open and close the stomate. When the stomates are open, they are the chief path through which gases,such as water vapour, corbon dioxide and oxygen, move from the leaf into the air and *vice versa*. Stomates may be present in both the upper and lower epidermis, but the number in the lower epidermis is usually greater than in the upper. Many, perhaps most, woody plants, have no stomates in the upper epidermis.

Mesophyll

Within the leaf, photosynthesis is localized in the mesophyll. This tissue, with the exception of the vascular bundles, comprises all the cells between the upper and lower epidermis. The cells of the mesophyll are thin-walled and when mature retain living protoplasm and a nucleus. Such thin-walled cells make up parenchyma tissue similar to the cells forming the softer parts of roots, stems, flowers, and fruits. The mesophyll is usually divided into two parts.

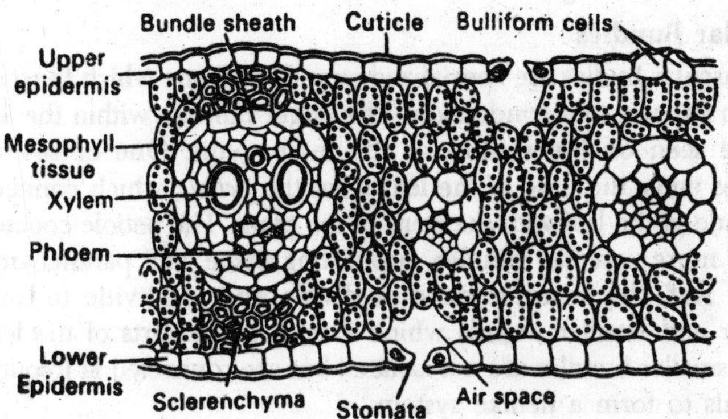

Fig. 2.2. Detailed structure of a part of T.S. of maize leaf.

The cells toward the upper side of the leaf are elongated at right angles to the surface and form one to three layers. These cells make up the *palisade* parenchyma, so called because of the resemblance of these cells to a palisade, or row of stakes forming a wall. Below the palisade cells, and extending to the lower epidermis, is a zone of a irregularly shaped cells with large intercellular spaces. These cells form the *spongy parenchyma.* The intercellular spaces of this tissue, together with similar but smaller spaces between the palisade cells, form a system of air passages extending throughout the leaf. All the mesophyll cells contain chloroplasts, embedded in the cytoplasm surrounding a large central vacuole.

The chloroplasts are most numerous in the palisade cells, perhaps two to three times as many as in the spongy tissue. In the sunflower leaf, for example, the cells of the palisade layer average 77 chloroplasts per cell, those of the spongy mesophyll 27 per cell. Estimating about 400,000 chloroplasts to each square millimeter of an elm leaf–a very conservative appraisal–it has been calculated that the total exposed area of the chloraplasts in the leaves of a mature elm tree may amount to as much as 140 square miles. In some plants, such as corn and other grasses, many aquatic plants, and conifers, the mesophyll is undifferentiated or only slightly differentiated into palisade and spongy tissue. Parenchyma cells containing chloroplasts are found in other parts of the plant, such as herbaceous stems, young twigs, and unripe fruits. The terms *chlorenchyma* is applied to such cells, whether they are found in the leaf or in other organs.

Vascular Bundles

Vascular budles are specialized strands of tissue which function both in support and conduction. The larger bundles within the leaf can be seen on the surface as the main veins. One or several bundles enter the base of the leaf from the *petiole*, which connects the blade of the leaf with the stem of the plant. The petiole contains one or more vascular bundles, supporting tissue, and parenchyma. In the leaf, the vascular bundles divide and subdivide to form smaller and smaller strands which extend to all parts of the leaf. These smaller bundles are not isolated but are connected at frequent intervals to form a netlike system.

In most cross-section of the leaf, therefore, vascular bundles may be seen both in end and in side view. The vascular bundles are usually located about halfway between the upper and lower epidermis. They are composed of two fundamental kinds of tissue, the *xylem* and *phloem*. Together these constitute the *vascular tissues* found in roots, stems, leaves, flowers, and fruits. The xylem and phloem are composed of elongated cells adapted to the movement of materials through the plants. The thick-walled xylem conducts water and supplies mechanical support.

The phloem is composed chiefly of thinner-walled cells and is the channel through which soluble foods are conducted. The xylem of flowering plants is composed of several kinds of cells. The most important, from the standpoint of conduction, is the *vessel element*, a somewhat elongated cell with open ends. A continuous series of such cells is termed a *vessel*. In some vessel elements the secondary wall is incomplete and is laid down in a pattern of rings or spirals inside the primary wall.

When mature, the vessel element is nonliving and consists only of the cell wall. The most important conducting cells of the phloem are the *sieve-tube elements*, so named because of the presence of sievelike openings in the end and frequently the side walls. Sieve-tube elements in the flowering plants are, like vessel elements, arranged end to end forming *sieve tubes*. The smaller vascular

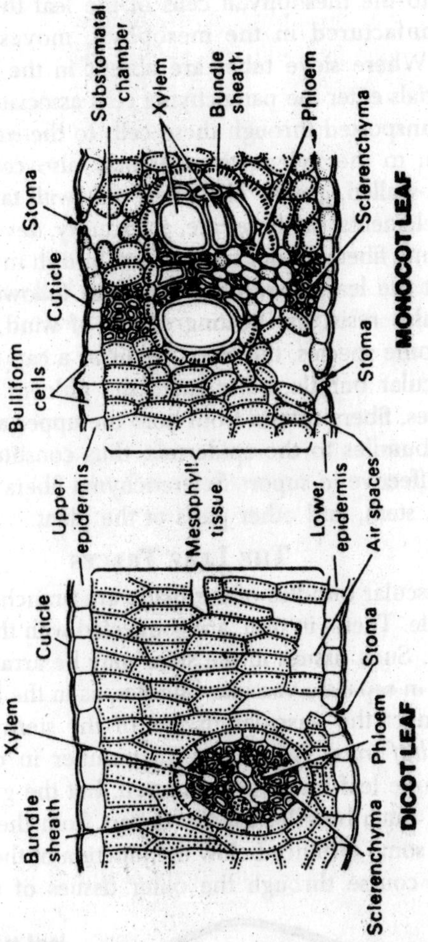

Fig. 2.3. Comparison of T.S. of a dicot and a monocot leaf.

bundles of the leaf are commonly surrounded by a sheath of thin-walled parenchyma. The cells of this sheath are in direct contact with both palisade and spongy tissue and probably facilitate the movement of materials between the vascular bundles and the mesophyll. They may also serve as a temporary storage tissue for reserve foods. The vascular bundles end blindly in the tissues of the leaf. As they terminate, they consist only of one or two xylem cells, together with a few parenchyma cells. Sieve tubes are absent.

So numerous are the smaller vascular branches that most cells of the mesophyll are only a few cells removed from conducting tissue. Water and mineral salts from the soil are continuously delivered to the mesophyull cells of the leaf through the xylem; food, manufactured in the mesophyll, moves in the opposite direction. Where sieve tubes are absent in the bundle ends, the food materials enter the parenchyma cells associated with the xylem and are transported through these cells to the nearest sieve tubes. In addition to the xylem, the leaf may also contain *sclerenchyma fibres*, thick-walled, greatly elongated cells with tapering ends. Like the vessel elements, the fibres are, at maturity, devoid of protoplasm. Sclerenchyma fibers do not conduct but furnish mechanical support. They assist the leaf to withstand collapse following loss of water, and they also resist the beating effects of wind and rain. In the leaves of some species, the fibers occur as a cap on the lower side of the vascular bundle, adjacent to the phloem. In the leaves of other species, fibers extend from both the upper and lower sides of the larger bundles to the epidermis, thus constituting a girderlike structure, effective in suport. Sclerenchyma fibers are also found in the petiole, stem, and other parts of the plant.

The Leaf Traces

The vascular bundles of the blade are branches of the bundles of the petiole. These, in turn, are connected with the vascular tissues of the stem. Such tissues in the stem may be arranged in a hallow cylinder or in separate vascular bundles, as in the leaf. The bundles which connect the vascular tissues of the stem and petiole are known as *leaf traces.* They vary in number in different species. One and three leaf traces are common, but the grasses and palms have many. Such bundles usually depart from the vascular system of the stem some distance below the junction of the leaf and, taking an upward course through the outer tissues of the stem, finally

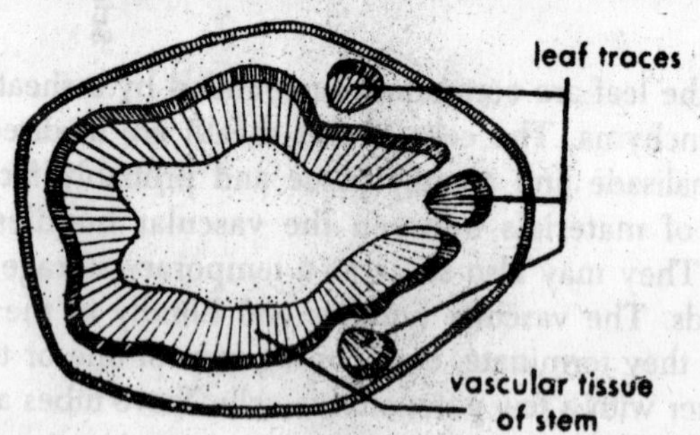

Fig. 2.4. Diagram of cross-section of red oak twig through the nodal region, showing the leaf traces.

enter the petiole. The leaf traces are the pathway along which water and other materials enter the leaf and foods pass from the leaf into the stem. In woody plants the leaf traces appear as bundle scars within the leaf scars on the twig, following the fall of the leaf.

Leaf Abscission

The falling of leaves is characteristics of woody plants. The fall, or *abscission*, of the leaf is related to the formation of a separation layer, a zone of specialized parenchyma cells extending across the base of the petiole. Toward the end of the growing season, the pectins of the cell walls of this layer dissolve through enzymatic action. The cells then separate and the petiole remains attached chiefly by the vascular bundles. The fall of the leaf is hastened by environmental factors, such as the shrinkage of the petiole on warm, sunny days, the impact of rain upon the leaf, or the formation of ice crystal within the separation layer. Before the fall of the leaf, or shortly afterwards, a protective layer of cork, which prevents the drying out of the exposed tissues, forms just below the separation layer.

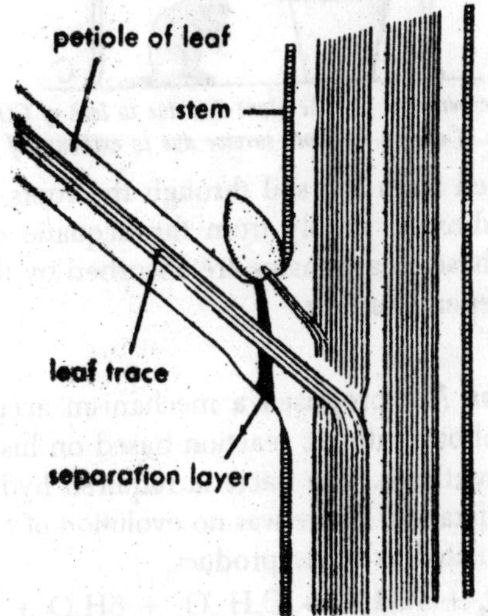

Fig. 2.5. Diagram showing position of separation layer.

Raw Materials

Carbon dioxide

In terrestrial plants, the carbon dioxide is obtained from the atmosphere through the stomata. A small amount of carbonates

Fig. 2.6. Priestley's experiment. A–The plant dies due to lack of CO_2. B–The mouse dies due to lack of oxygen. C–Both survive due to exchange of gases.

are also absorbed from the soil through the roots. Aquatic plants obtain carbon dioxide supply from the acquatic environment as bicarbonates. These bicarbonates are absorbed by the hydrophytes through their general surface.

Water

In 1924, *Van Niel* proposed a mechanism accounting for the dual nature of photosynthetic reaction based on his observation of bacterial photosynthesis. The bacteria required hydrogen sulphide for their carbon fixation. There was no evolution of oxygen. Sulphur globules accumulated as waste product.

$$6CO_2 + 12H_2S \rightarrow C_6H_{12}O_6 + 6H_2O + 12S$$

Van Niel suggested that water was a substance from which H^+ ions and electrons could be derived for the reduction of carbon dioxide in oxygenic photosynthesis of all organisms.

$$2H_2O \xrightarrow{\text{light}} O_2 + 4(H)$$

$$4(H) + CO_2 \longrightarrow CH_2O + H_2O$$

The nature of the reductant was not known, and it was clear that the second reaction was complex and probably involved several steps.

Ruben and *Kamen* (1941) were able to use isotopically enriched oxygen (O_2) to established that all of the oxygen produced in photosynthesis was derived from water and none from carbon dioxide.

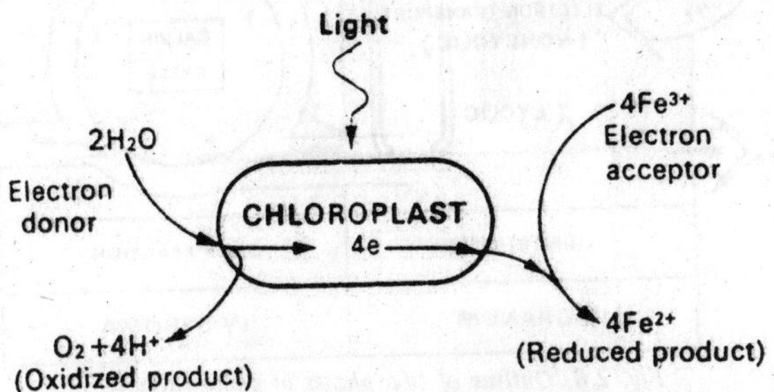

Fig. 2.7. Outline of Hill reaction.

Hill (1937) illuminated the isolated chloroplasts of *Stellarta media* in the presence of leaf extract or hydrogen acceptors in the absence of carbon dioxide. The chloroplast evolved oxygen. The reaction involving the production of oxygen by the illuminated chloroplasts in the absence of carbon dioxide fixation is called *Hill reaction.* The hydrogen acceptor of Hill reaction is nicotinamide adenine dinucleotide phosphate (NADP).

$$2NADP^+ + 2H_2O \xrightarrow[\text{light}]{\text{chlorophyll}} 2NADPH + 2H^+ + O_2$$

Light

Since the primary reaction of photosynthesis involve the absorption of light by pigments. It is necessary to examine some of its basic properties. Light is electromagnetic energy propagated in discrete corpuscles called *photons* or *quanta.* As the energetics of chemical reactions are usually described in terms of kilocalories per molecule of the chemicals (1 mole = 6.02×10^{23} quanta). The colour of the light is determined by the wavelength (λ) of light radiation. At any given wavelength , all the quanta have the same energy. The energy (E) of a quantum is inversely proportional to its wavelength.

Depending upon the wavelength electromagnetic spectrum consists of eight types of radiations-cosmic rays, gamma rays, X-

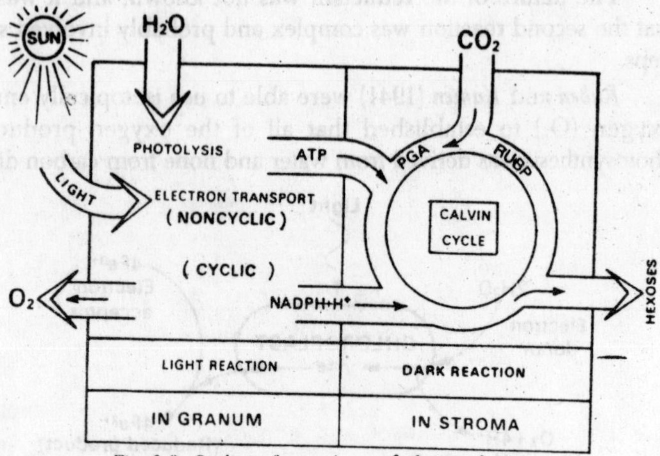

Fig. 2.8. Outline of two phases of photosynthesis.

rays, ultra-violet rays, light spectrum, infrared rays, electric rays and radio rays. Visible light consists of radiations having a wavelength between 390-760 nm. It can be resolved into light of different colours-*violet* (390-430 nm), *blue* (430-470 nm), *bluegreen* (470-500 nm), *green* (500-580 nm), *yellow* (580-600 nm), *orange* (600-650 nm), *orange-red* (650-660 nm) and *red* (660-760 nm). Red light above 700 nm is called *infrared. Infrared rays* having wavelength of 760-10,000 nm. Radiation shorter than those of violet are called *ultra violet rays* having a wavelength of 100-390 nm. Solar radiations or sunlight reaching the earth have wavelength between 300-2600

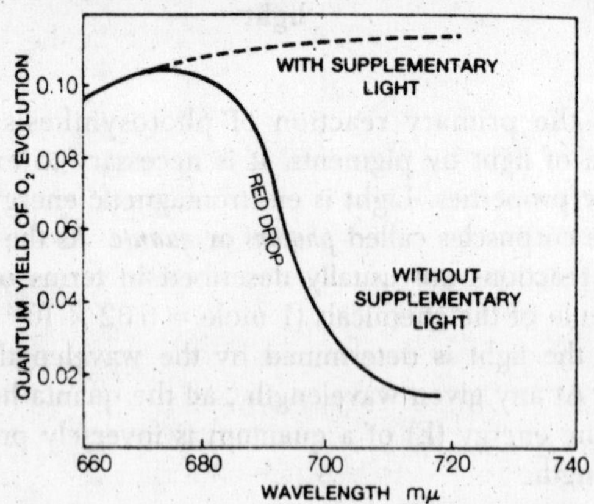

Fig. 2.9. Red drop and the Emerson's effect on Chlorella.

nm. The proportion of light energy that is available in photosynthesis has long been a matter of controversy.

Much attention was focused in the period 1930-1950 on the quantum requirement of photosyntehsis, a measure of the efficiency of the process. The initial thought was that, since 1 quantum can conduct only one molecular reaction, the number of quanta absrobed per molecule of oxygen produced might indicate the number of reaction steps involved. In complex experiments *Warburg* found quantum requirements of 4 quanta per molecule of oxygen produced. This means 1 quantum per (H) derived from H_2O for the reduction of carbondioxide. The energy required to derive one reducing equivalent from H_2O is close to 30 kcal/mole, and the energy available in quanta of red light, which *Warburg* used in his experiments is only about 40 kcal/einstein.

Photosynthetic Pigments

The photosynthetic pigments are those which occur inside the chloroplasts and can take part in absorption of light energy for the purpose of photosynthesis. There are two catagories of these pigments *chlorophyll* and *carotenoids.*

Chlorophyll

So far the only light- absrobing pigment we have mentioned is chlorophyll. Initially the green pigment of plants was recognized as the substance responsible for light absorption in photosynthesis, absorbing red and blue light but not green. There are five types of chlorophyll occurring in plants. These are chlorophyll *a, b, c, d* and *e.* Bacteria possess two types of related pigments. These are *bacteriochlorophyll* and *bacterioviridin.* In higher plants only chlorophyll *a* and *b* are present. Chlorophyll *a* occurs in all the photosynthetic plants. It is bluish green in the pure state. Its imperical formula is $C_{55}H_{72}O_5N_4Mg$. Chlorophyll *b* is olive green in pure state and its empirical formula is $C_{55}H_{70}O_5N_4Mg$.

Both the chlorophylls are soluble in fat solvents such as ether, chloroform, carbon tetrachloride, alcohols etc. Chlorophyll *a* is more soluble in ether whereas chlorophyll *b* in methyl alcohol. In 1912, the structure of chlorophyll was studied by *Wilstatter*, *Stoll* and *Fischer.* The chlorophyll is a tetrapyrrole and bears a close resemblance to the chemical structure of heme and the cytochromes. Chlorophyll differs from these iron enzymes in that it contains an atom of magnesium, which does not appear to participate directly

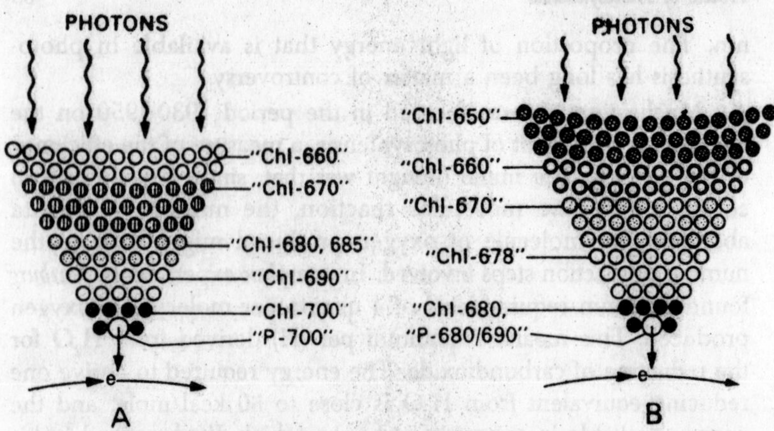

Fig. 2.10. Schematic representation of harvest of light energy by pigment system I (A) and by pigment system II (B).

in electron transfer reactions as does the iron of cytochromes. The chlorophyll has a tadpole-like structure, with a head called *porphyrin* and a *tail* (made up of long chain alcohol called *phytol*). The head is made up of four *pyrrole rings* which are linked by methane bridges (–CH=). The skeleton of pyrrole ring is made up of 5 atoms–4 carbon and one nitrogen. The nitrogen lies towards the centre. A megnisium atom is held in the centre of porphyrin head by nitrogen atoms of pyrrole rings (with the help of 2 convalent and 2 co-ordinate bonds).

The external carbon atoms of the pyrrole rings have been given specific numbers, 1-8 carbon atoms 1,3,5 and 8 have methyl groups ($-CH_3$). Carbon atom 2 possesses a vinyl group ($-CH=CH_2$) while carbon atom 4 has an ethyl group ($= CH_2-CH_3$). Carbon atom 6 is attached to methane group by an acetoacetic group. Carbon atom 7 is connected to the phytol tail through a propionic acid residue. Chlorophyll *b* differs from chlorophyll *a* in having formyl group (–CHO) instead of methyl ($-CH_3$) group at carbon atom 3. Chlorophyll *b* is present in most green plants, but its place is taken by *phycocyanin* blue-green aglae, *fucoxanthin* in brown algae and *phycoerythrin* in red-algae. These are called *accessory phycobilin pigments.* The chemical structure of these accessroy phycobilin pigments is similar to that of chlorophyll in that they are tetrapyrrole compounds; however, they differ in being linear instead of cyclic in structure, and they do not have a metal component.

Carotenoids

The carotenoids are a group of yellow, brownish is reddish pigments which are associated with the chlorophyll inside the chloroplast. These remain alone inside the chromoplasts. Like chlorophyll *b* these are also called accessory photosynthetic pigments. They hand over the energy to chlorophyll *a* absorbed by them. The carotenoids are of two types:

(a) *Carotenes.* These are closely related to the vitamin A and are basically long chain lipophilic molecules with a more active terminal group at each end. The empirical formula is $C_{40}H_{56}O$ i.e., cryptoxanthin and $C_{40}H_{56}O_2$ i.e., zeaxanthin.

The carotenoids are soluble in fat solvents like ether, chloroform, acetone, carbon disulphide.

(b) *Fluorescence.* The fluorescence is the property of emmission of radiations absorbed by a substance and the substance which can emit back the absorbed radiations is called *fluroscent substance.* It has been noted that all the photosynthetic pigments are fluroscent substances. However, most of the fluroscence emitted by photosynthetic organs is due to chlorophyll *a* because other pigments hand over their absorbed energy to it through resonance.

The normal state of the atoms or molecules is called *ground state* or *singlet state,* when an electron of a molecule or atom absorbs a quantum of light, it is raised to a higher energy level which is called *excited second singlet state.* The excited state is unstable and lasts for about 10^{-9} second. The exicited electron may return to ground state in two ways (a) either losing its remaining energy in the form of heat or (b), by losing extra energy in the form of radiation energy. This later process is called fluorescnce. The fluroscent substances emit fluorescence. The fluroscent substances emit fluroscent light only during the period they are exposed to incident light. The fluroscent light is of higher wavelength than the incident light.

Absorption Spectrum of Chlorophyll

If a substance is extracted and illuminated with light of various wavelengths, it is possible to dertermine its absorption spectrum, the amount of light of different wavelengths that the substance absorbs or does not absorb. Chlorohyll's absorption spectrum has peaks in the blue and red regions, which means that light of these wavelengths in absorbed. The major peak at 675 nm and minor peak at 450nm. The unabsorbed or reflected light appears green, thus making chlorophyull, the chloroplasts and the leaves that contian it, appear green to our eye.

Action Spectrum

The curve showing the relative rates of photosynthesis at different wavelengths of light is called action spectrum. The curve shows that maximum photosynthesis occurs in red part and blue-violet part of light. Thus the action spectrum of photosynthesis corresponds closely to absorption spectra of chlorophyll *a* and *b* showing the chlorophyll *a* as well as *b* are the main photosynthetic pigments, however, photosynthesis also occurs in the midpart of light spectrum where carotenoid are active.

Photosynthetic Units (PSU)-Quantosome

As already stated that light travels in the form of waves of different wavelengths. In certain respects light behaves as if it consists of tiny particles called *photons* or *quanta.* The quanta is indivisible units. Using the technique of intermittent light in their experiments, *Emerson* and *Arnold* (1932) found that the evolution of one molecule of oxygen by *Chlorella* required the functioning of approximately 2500 molecules of chlorophyll. It was also estimated that in reduction of one molecule of carbon dioxide and the liberation of one molecule of oxygen eight quanta of light energy are consumed. Putting these observations together it is easy to deduce that a group of 300 pigment molecules is the smallest unit that is capable of utilising one quantum of light energy. This is called the *photosynthetic unit.*

Under electron microscope made by *Park* and *Biggins* (1964) showed that minute structures about 100 Å in diameter are attached to the chloroplast lamellae. These strips of lamellae contained about 230 chlorophyll molecules each and represented photosynthetic units to which they gave the name of *quantosome.* The quantosome has a *photocentre* or *reaction centre* which is fed by about 200 harvesting pigment molecules. The photocentre consists of a special cholorohyll *a* molecule. It absorbs light energy at longer wavelengths.

The harvesting molecules are of two types, the *antenna molecules* and *core molecules.* The antenna molecules get excited by absorbing light of various wavelengths but shorter than that of trap centre. The antenna molecules come to ground state by hand over their energy to the core molecules. The energy picked up by core molecules is supplied to the reaction centre. On absorption of this energy, the reaction centre gets excited and extrudes an electron after which it comes to ground state to repeat the cycle.

Photosystems or Pigment Systems

Photosynthetic units occur in the form of two distinct groups called *photosystems* I and II or *pigment systems* I and II. In pigments system, I, there are different types of molecules of chlorophyll *a* which absorb photons of light, on the basis of type of wavelength of light absorbed by these chlorophyll *a* molecules they are known as *chl-660, chl-670, chl680* and *chl 690.* At the end of this energy gathering pigment system is present a special type of chlorophyll *a* molecule called *P-700* which absorbs 700 nm of wavelength of light which functions as reaction centre. In *P-700* photochemical reaction takes place.

Pigment system I has a reducing agent Fe 5 centre A, Fe 5 centre B (ferrodoxine), plastoquinone, cytochrome complex and plastocyanin. Normally it derives an electron from pigment system II to $NADP^+$ for cyclic photophosphorylation independently. The pigment system I is located on both the non-appressed part of grana thylakoids and stromal thylakoids. In pigment system II, the photocentre is a special chlorophyll *a* molecule called P-680. Besides it contains *chl-650, chl-660, chl-670, chl-678.* It is surrounded by chlorophyll *b* and carotenoid molecules. Pigment system II also contains Mg^{++}, Cl^-, Quencher molecule Q, plastoquinone, cytochrome complex and plastocyanin. It picks up electrons released during photolysis of water. In this system photons of lower wavelength of light are absorbed. It is inactive infrared light. This system is located in the appressed part of grana thylakoids only.

Mechanism of Photosynthesis

Photosynthesis is an oxidation-reduction process, in which water is oxidized to release oxygen and carbon dioxide is reduced to form carbohydrates. Photosynthesis occurs in two phases, ***photochemical reactions*** (also called ***light*** or ***Hill reaction***) and ***biochemical reactions*** (also called *dark* or *Blackman's* reaction).

The Light or Photochemical Reaction in Photosynthesis

Evidences for Light Reaction

Following evidences indicate that the over all reaction in photosynthesis takes place in two steps: one is light mediated and the other is light independent.

Temperature coefficient studies

The rate of photosynthesis of two groups of plants were compared. Both were supplied with an adequate concentration of carbon dioxide, but one group was kept under light of high intensity and other group in light of low intensity. When the rates of photosynthesis were measured at different temperatures. It was found that *high light* group had a Q_{10}=2 but the *low light* group Q_{10}=1. Strictly, chemical reactions characteristically have a Q_{10} from 2 to 3. This fact indicates that at least one of the reaction involved in photosynthesis is of a purely chemical type. Since it was first pointed out by *Blackman*, this reaction is called as dark reaction. The other reaction of Q_{10} indicated that one of the reactions proceeds only at the expanse of absorbed light. It is called dark reaction. The Q_{10} of light reaction is I.

Flashing light experiments

Photosynthesis involves photochemical and biochemical reactions is also shown by the results of investigations in which plants are exposed to intermittent light. *Warburg* (1919) obtained higher rates of photosynthesis in *Chlorella* when it was exposed to rapid and alternate periods of light and darkness instead of continuous illumination. *Emerson* and *Arnold* (1932) found that at 25°C, the maximum photosynthesis took place when light and dark periods were respectively 10^{-5} second and 0.055 second. At 1.1°C maximum photosynthesis could be obtained with the same light period but the dark period had to be increased to same light period but the dark period had to be increased to 0.4 second. It means the temperature influences reactions of the dark period and reactions of the light phase are photochemical.

Operation of photosystem II

The main function of PS II is the photolysis of water to release oxygen. In this act, when pigments of PS II absorb light , electrons are released from this system, in this manner chlorophyll molecules get oxidized. Oxidized pigment then take electrons from OH^- ions and are again reduced. The OH (hydroxyl radical) is left from OH- ions and are again reduced. The OH (hydroxyl radical) is left from OH^- and 4(OH) radicals give rise to two molecules of water and one moleclue of oxygen. In the above process H^+ atoms are released which are used for the reduction of $NADP^+$. Manganese ions play an important role in this process.

$$
\begin{array}{lcl}
4H_2O & \rightarrow & 4\ H^+ + 4OH^- \\
4OH & \rightarrow & 4(OH) + 4e^- \\
4(OH) & \rightarrow & 2H_2O + O_2 \\
\hline
2H_2O & \rightarrow & O_2 + 4e^- + 4H^+
\end{array}
$$

Electrons released in PS II reduced quencher molecule Q which has a potential (E'o) of about 0-01 V. From reduced Q the electrons

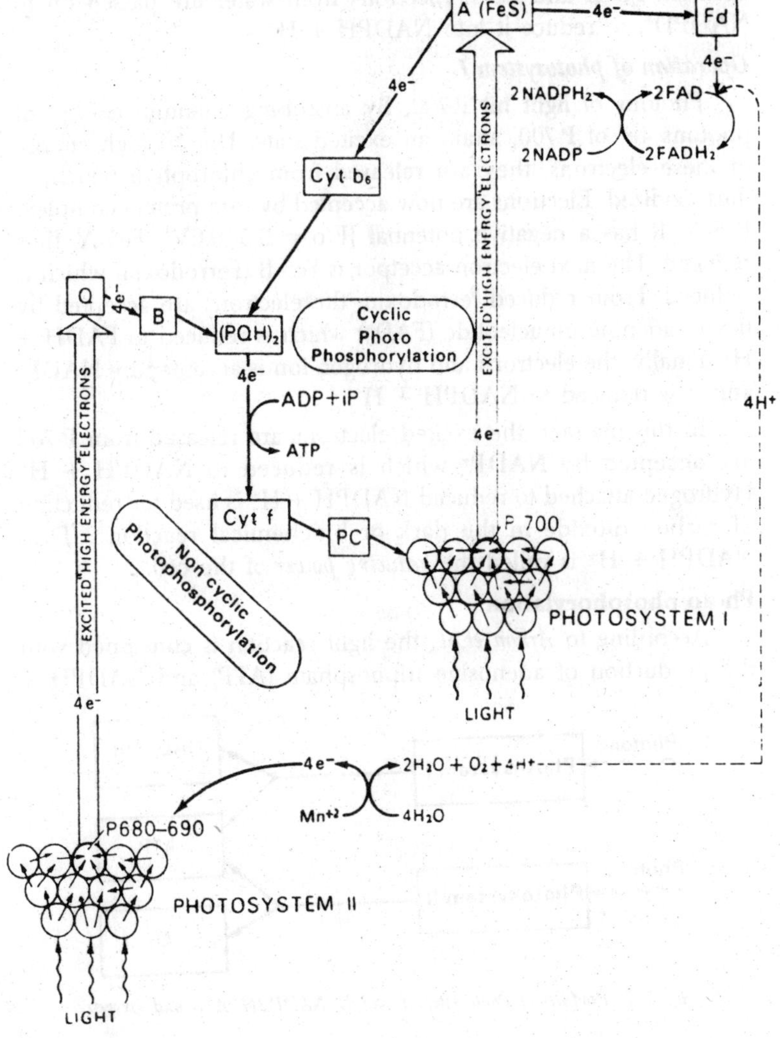

Fig. 2.11. Generalized scheme of two light reactions in photosynthesis.

are accepted by B (an unknown substance) which is reduced. From reduced B the electrons are accepted by plastoquinone (PQ). Electrons from reduced PQ are passed to the cytochrome *f* which is now reduced. Again the electrons from reduced cytochrome *f* are accetped by plastocyanin, (a copper enzyme that has an E'o close to 0.4V), which is thus reduced. Finally, the electrons from reduced plastocyanin are accetped by P-700*f* photosystem I. P-700 is now reduced. In this way plastocyanin works as a middle man between PS II and PS I. Electrons from water are passed on to $NADPD^+$, to reduce it into NADPH + H^+.

Operation of photosystem I

Photons of light hit P-700, By absrobing quantum energy of photons $4e^-$ of P-700, attain an excited state. Due to high energy of these electrons, they are released from chlorophyll which, is thus oxidized. Electrons are now accepted by iron-protein complex, Fe5-A. It has a negative potential (E'o = 0.5–0.6V). Fe5-A. thus reduced. The next electron accetpor is Fe5-B (Ferrodoxin) which is reduced. From reduced ferrodoxin, the electrons are accepted by flavin -adenine dinucleotide (FAD^+) which is reduced to FADH + H^+. Finally, the electrons and hydrogen ion is accepted by $NADP^+$ and it is reduced to NADPH + H^+.

In this manner, the excited electrons are released from P-700 are accepted by $NADP^+$ which is reduced to NADPH + H^+ Hydrogen attached to reduced NADPH + H^+ is used for reduction of carbon dioxide in the dark or biochemical reaction . Thus NADPH + H^+ is called the *reducing power* of the cell.

Photo-phosphorylation

According to *Arnon et.al.*, the light reaction is concerned with the production of adenosine triphosphate (ATP) and NADPH +

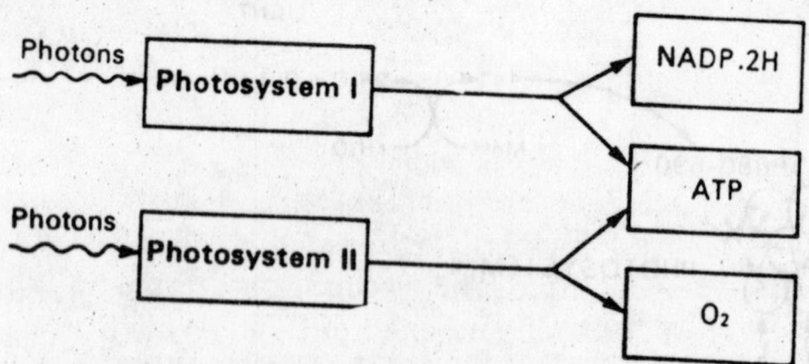

Fig. 2.12. Products of Photosystem I and II–NADP.2H, ATP and Oxygen.

H+, which is the reduced form of nicotinamide adenine dinucleotide phosphate. ATP is a high energy compound and is formed by the addition of an inorganic phosphate group to adenine diphosphate (ADP) by a process called photophosphorylation.

Photophosphorylation takes place in two ways :

Cyclic photo-phosphorylation

It takes placce most efficiently when the Hill reaction is prevented, i.e., the medium is devoid of the hydrogen acceptor NADP. Here the electrons released from P-700 move through Fe5-A to cytochrome *b6,* to plastoquinone to cytochrome-*f* to plastocyanin and again return to P-700. While passing between ferrodoxin and plastoquinone, the electrons lose sufficient energy to form ATP from ADP and inorganic phosphate (P_i)

$$ADP + P_i \xrightarrow[\text{Chlorophyll}]{\text{Light}} ATP$$

well-known poisons of photosynthesis 3-(3, 4 dichlorophenyl) dimethyl urea (DCMU) do not inhibit the cyclic photophosphory-lation.

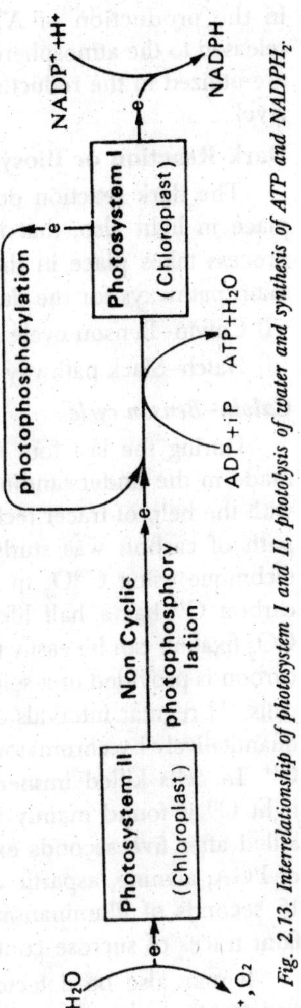

Fig. 2.13. Interrelationship of photosystem I and II, photolysis of water and synthesis of ATP and NADPH₂.

Non-cyclic photophosphorylation

It is the normal process of photophosphorylation and is unidirectional. It is carried out in collaboration of both photosystems I and II. Photoexcited chlorophyll molecules in photosystem II donate electrons to acceptors. The extruded electron has an energy equivalent to 23 kcal/mole. The electrons pass through Q, B, PQ, cytochrome-*f*, plastocyanin and enter of P-700. While passing over plastoquinone to cytochrome-*f*, the electron loses sufficient energy for the synthesis of ATP. The overall photochemical reaction results in the production of ATP and reduced NADP and oxygen is released to the atmosphere. The products of photochemical reaction are utilized in the reduction of carbon dioxide to the carbohydrate level.

Dark Reaction or Biosynthetic Phase or Blackman's Reaction

The dark reaction does not require light, although this takes place in light also, that is why this is called dark reaction. This process takes place in the stroma of chloroplasts. There are two main pathways for the dark reaction.

(i) Calvin–Benson cycle

(ii) Hatch–Slack pathway

Calvin–Benson cycle

During the last four decades considerable progress has been made in the understanding of the dark reaction in photosynthesis with the help of tracer techniques and paper chromatography. The path of carbon was studied with the help of radioactive tracer technique using $C^{14}O_2$ in a unicellular alga, *Chlorella.* Radioactive carbon C^{14} has a half life of 5200 years. Therefore, the path of CO_2 fixation can be easily traced with its help. Here the readioactive carbon is provided in a solution of bicarbonate to photosynthesizing cells. At regular intervals of time, the cells are killed and analyzed quantitatively by chromatography for various compounds contianing C^{14}. In cells killed immediately after one second expose to the light C^{14} is found mainly in phosphoglyceric acid (PGA). In cells killed after five seconds expose to the light, C^{14} appears in traces of PGA, alanine, aspartic acid and malic acid. In cells killed after 15 seconds of illumination, in addition to the above substances, faint traces of sucrose containing C^{14}.

It may also be detected in some pentose, glucose phosphate and fructose phosphate. After one minute of illumination, in the killed cells abundant C^{14} occurs in glucose phosphate and fructose phosphate. In lipids and proteins also, the C^{14} appears after 5 minutes of illumination. Thus radioactive C^{14} obtained from $C^{14}O_2$ is rapidly spread in all the compounds found in plant cell. After many panstaking calculations, *Calvin* worked out the pathway of carbon dioxide fixation. The acceptor molecule was found to be ribulose-1,5 biphosphate. The followings are the steps taking place in dark reaction of photosynthesis.

1. *Phosphorylation I.* Carbon dioxide reduction starts with 5-carbon sugar i.e., ribulose monophosphate. Six molecules of ribulose monophosphate react with 6 molecules of ATP (produced during light reaction) to form 6 molecules of *ribulose* 1, 5 *biphosphate* and 6 molecules of ADP.

$$\text{6 Ribulose-5-phosphate} + \text{6 ATP} \xrightarrow{\text{Phosphopentokinase}} \text{6 Ribulose- 1,5 biphosphate} + \text{6 ADP}$$

2. *Carboxylation.* Ribulose 1, 5 biphosphate (RuBP) acts as CO_2 acceptor. 6 molecules of RuBP react with 6 molecules of CO_2 and 6 molecules of water giving rise 12 molecules of 3-phosphoglyceric acid.

$$\text{6 Ribulose-1, 5 biphosphate} + 6\ CO_2 + 6H_2O \xrightarrow[\text{Carboxylase}]{\text{Ribulose 1, 5 biphosphate}} \text{3-phosphoglyceric acid (12 molecule)}$$

3. *Phosphorylation II.* 12 molecules of 3- phosphoglyceric acid react with 12 molecules of ATP giving rise to 12 molecules of 1,3-diphosphoglyceric acid and 12 molecules of ADP.

$$\text{3-phosphoglyceric acid} + \text{12 ATP (12 molecules)} \xrightarrow[\text{Kinase}]{\text{Phosphoglyceric}} \text{1, 3-diphosphoglyceric acid} + \text{12 ADP (12 molecule)}$$

4. *Reduction.* 12 molecules of NADPH + H^+ produced in light reaction are now used to reduces 12 molecules of 1, 3-diphosphoglyceric acid leading to the formation of 12 molecules each of 3- phosphoglyceraldehyde or glyceraldehyde-3-phosphate, $NADP^+$ and phosphoric acid.

$$\text{1, 3 diphosphoglyceric acid} + \text{12 NADPH} + H^+ \text{(12 molecules)} \xrightarrow[\text{dehydrogenase}]{\text{Triose phosphate}} \text{3-phosphoglyceraldehyde} + 12\ NADP^+\ 12\ H_3PO_4 \text{ (12 molecules)}$$

In this way by the reduction of carbon dioxide 12 molecules of 3-phosphoglyceraldehyde are formed. Out of these 12 molecules only two molecules go to synthesize sugar and test 10 molecules are recycled to produce 6 molecules of ribulose 5-phosphate.

5. One molecule of 3-phosphoglyceraldehyde isomerize into dihydroxyacetone phosphate.

$$\text{3-phosphoglyceraldehyde} \underset{\text{isomerase}}{\overset{\text{Triose phosphate}}{\rightleftharpoons}} \text{3-dihydroxyacetone phosphate}$$

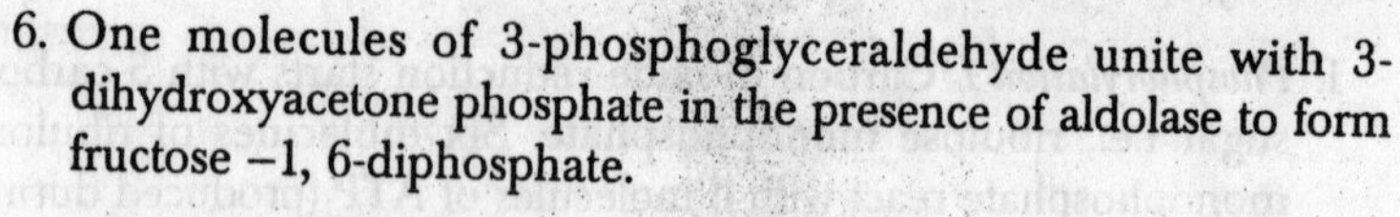

6. One molecules of 3-phosphoglyceraldehyde unite with 3-dihydroxyacetone phosphate in the presence of aldolase to form fructose –1, 6-diphosphate.

3-phosphoglyceraldehyde + 3-dihydroxyacetone phosphate $\xrightarrow{\text{Aldolase}}$ Fructose-1, 6-diphosphate.

7. Three molecules of phosphoric acid are released from one molecule of fructose-1,6-diphosphate to form one molecule fructose-6-phosphate in the presence of phosphatase.

Fructose-1, 6 diphosphate + H_2O $\xrightarrow{\text{Phosphatase}}$ Fructose-6-phosphate + $3H_2PO_4$.

8. Fructose -6-phosphate is converted to fructose-1-phosphate which forms glucose-1-phosphate.

Fructose-6-phosphate $\longrightarrow$ fructose -1-phosphate

fructose -1-phosphate $\longrightarrow$ Glucose -1-phosphate

9. Fructose -1-phosphate and glucose-1- phosphate condense to form sucrose.

Fructose-6-phosphate + Glucose-1- phosphate $\xrightarrow{\text{Phosphatase}}$ sucrose.

10. *Regeneration of ribulose-5-phosphate.* Remaining 10 molecules of 3-phosphoglyceraldehyde regenerate 6 molecues of ribolose-6-phosphate after a series of complex reactions as follows.

(a) 3-phosphoglyceraldehyde reacts with fructose-6-phosphate to yield a pentose (5-carbon) xylulose-5-phosphate and a triose (4-carbon) erythrose-4- phosphate in the presence of transketolase.

Fructose-6-phosphate + 3-phosphoglyceraldehyde $\xrightarrow{\text{Transketolase}}$ Xylulose-5-phosphate + Erythrose-4-phosphate.

Xylulose-5- phosphate readily isomerizes to ribulose-5-phosphate.

(b) Erythrose-4-phosphate combines with dihydroxyacetone phosphate in the presence of aldolase to form sedoheptulose 1,7-diphosphate, a 7-carbon sugar.

Erythrose-4-phosphate + Dihydroxyacetone phosphate $\xrightarrow{\text{Aldolase}}$ Sedoheptulose

(c) Sedoheptulose-1, 7-diphosphate looses one phosphate group in the presence of phosphotase to form sedoheptulose -7-phosphate.

$$\text{Sedoheptulose-1, 7, diphosphate} \xrightarrow{\text{Phosphatase}} \text{Sedoheptulose-7-phosphate}$$

(d) Sedoheptulose -7-phosphate reacts with 3-phosphoglyceraldehyde in the presence of transketolase to form xylulose-5-phosphate and ribose-5-phosphate.

$$\text{Sedoheptulose-7-phosphate + 3-phosphoglyceraldehyde} \xrightarrow{\text{transketolase}} \text{Xylulose-5-phosphate + ribose-5-phosphate.}$$

(e) Xylulose-5- phosphate is converted into ribulose-5- phosphate(5-carbon sugar) in the presence of phosphoketopentose epimerase.

$$\text{Xylulose-5-phosphate} \xrightarrow[\text{epimerase}]{\text{Phosphoketopentose}} \text{Ribulose-5-phosphate}$$

(f) Ribose-5-phosphate is also converted in ribulose-5- phosphate in the presence of phosphopentose isomerase.

$$\text{Ribose-5-phosphate} \xrightarrow[\text{isomerase}]{\text{Phosphopentose}} \text{Ribulose-5-phosphate}$$

(g) Ribulose-5-phosphate is finally converted into ribulose 1,5-diphosphate in the presence of phosphopnetose kinase and ATP.

$$\text{Ribulose-5-phosphate + ATP} \xrightarrow[\text{Kinase}]{\text{Phosphopentose}} \text{Ribulose-1, 5-diphosphate + ADP}$$

In this manner the Calvin's cycle is completed. The net reaction of C_3 dark fixation of CO_2 is

$$6RuBP+6CO_2+18ATP+12NADP_2H \rightarrow 6RuBP+12NADP^+ + C_6H_{12}O_6+18ADP+18P.$$

The Hatch-Slack cycle or C4 pathway

It was worked out by *Hatch* and *Slack* (1965,1967). *Kortschak et.al.* (1965) first of all found that labelled carbon dioxide ($C^{14}O_2$) assimilated by sugarcane leaves first appeared in 4-carbon compound i.e., oxaloacetic acid. *Hatch* and *Slack* found it a regular mode of carbon dioxide fixation in a number of tropical plants such as maize, Sorghum *Salsola, Atriplex, Amaranthus* etc. These plants are called C^4 plants because of the first stable photosynthetic product being a 4-carbon compound. These plants live in hot, arid and

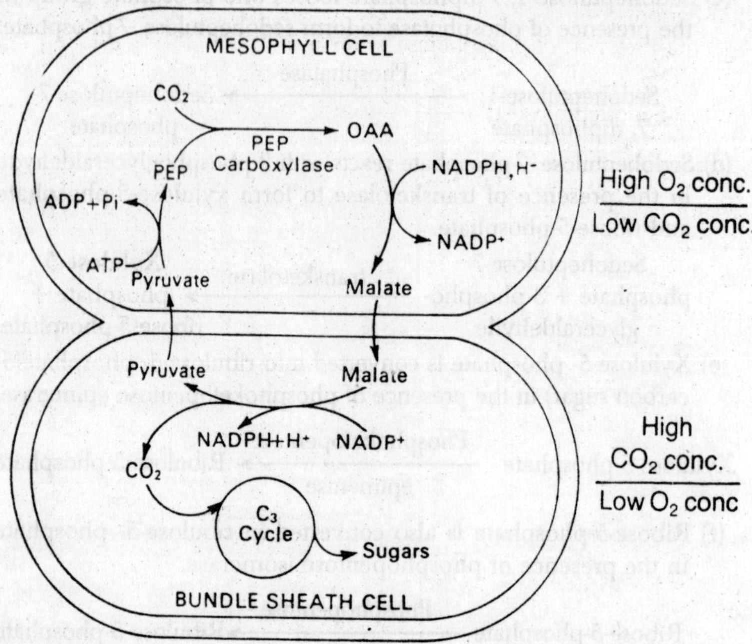

Fig. 2.14. Hatch-Slack pathway (Cycle).

saline habitats. These plants have Krang anatomy as the mesophyll is undifferentiated and its cells occur in concentric layers around vascular bundles. In C^4 plants, initial fixation of CO_2 occurs in

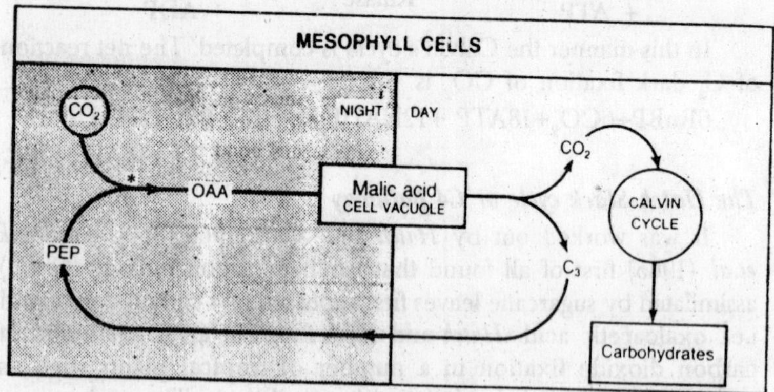

Fig. 2.15. At night the stomata are open and CO_2 is fixed by the action of PEP carboxylase (shown by asterisk) Malic acid is stored in the cell vacuole. By day the stomata are closed, CO_2 is released from Malic acid and fixed by Calvin cycle.

mesophyll cells. In this cycle, the primary acceptor of CO_2 is phosphoenol pyruvic acid. (PEP).

I-XII Enzymes: I, Carboxydismutase, II,2-phosphoglycerate kinase, III. NADP+ glyceraldehyde-3-phosphate dehydrogenase, IV. Triose phosphate isomerase, V. Vldolase VI Fructose-1,6-diphosphate, VII. Transketolase, VIII, Ribulose 5-phosphate-3-epimerase, IX. Aldolase, X, Sedoheptulose-1,7- disphosphate, XI, Transketolase, XII. Ribosephosphate isomerase. XIII, Phosphoribulokinase.

1. The first step involves the carboxylation in phosphoenol pyruvic acid in the chloroplasts of mesophyll cells to form oxaloacetic acid and phosphoric acid in the presence of phosphoenol pyruvic acid carboxylase.

$$\text{PEP} + CO_2 + H_2O \xrightarrow[\text{acid carboxylase}]{\text{Phosphoenol pyruvic}} \text{Oxaloacetic acid} + H_3PO_4$$

2. Oxaloacetic acid is reduced to malic acid or aminated to form aspartic acid in the presence of transaminase and malate dehydrogenase respectively.

$$\begin{matrix}\text{Oxaloacetic acid} + \\ NADPH_2\end{matrix} \xrightarrow{\text{malate dehydrogenase}} \begin{matrix}\text{Malic acid} + \\ NADP^+\end{matrix}$$

$$\begin{matrix}\text{Oxaloacetic + acid} + \\ NH_3 + NADPH_2\end{matrix} \xrightarrow{\text{transaminase}} \begin{matrix}\text{Aspartic acid} + \\ NADP^+ + H_2O\end{matrix}$$

3. From chloroplasts of mesohpyll cells,the malic acid or aspartic acid is translocated to the chloroplasts of bundle sheath cells where it is decarboxylated as in malic acid or deaminated as in aspartic acid to form pyruvic acid and carbon dioxide.

$$\begin{matrix}\text{Malic acid} + \\ \text{NADP} +\end{matrix} \xrightarrow{\text{decarboxylase}} \begin{matrix}\text{Pyruvic acid} + CO_2 \\ + NADPH_2.\end{matrix}$$

4. Now second carboxylation occurs in the chloroplasts of bundle sheath cells through Calvin cycle. RuBP of Calvin cycle is called *final* or *secondary* acceptor of CO_2 in C_4 plants. The pyruvic acid produced above is sent back to mesophyll cells. Here it is phosphorylated to regenerate phosphoenol pyruvic acid. Energy is required for this. It is provided by ATP.

$$\begin{matrix}\text{Pyruvic acid} + \\ P_i + ATP\end{matrix} \xrightarrow[\text{Kinase}]{\text{Pyruvate } P_i} \begin{matrix}\text{Phosphoenol pyruvic acid} + \\ AMP + PP_i \text{ (Pyrophosphate)}\end{matrix}$$

Ribulose-diphosphate accepts CO_2 produced in step 3 in the presence of carboxydimutase system and ultimately yields 3-phosphoglyceric acid as mentioned above in Calvin cycle. Some of the 3-phosphoglyceric acid is utilized in the formation of glucose-6-phosphate and sucrose while regenerated ribulose-1,5-biphosphate in the system. In the Calvin cycle 3 ATP molecule are required for the C^4 plants, 5 ATP molecules are required for the fixation of CO_2. For the formation of a glucose molecules C_4 plants require 30 ATP while C_3 plants utilize only 18 ATP.

6-phosphoenol pyruvic acid + 6RuBP + 30ATP + 12 $NADPH_2$ → 6 phosphoenol pyruvic acid + 6RuBP + $C_6H_{12}O_6$ + 30 ADP + $12NADP^+$ + $30H_3PO_4$.

Significance of Hatch-Slack pathway

(i) This pathway is a modification of Calvin cycle and its advantageous to plants growing in dense, stands of tropical vegetation where the carbon dioxide concentration may be very much reduced.

(ii) Concentric arrangement of mesophyll cell produce a smaller area in relation to volume for better utilization of available water and reduce the intensity of solar radiations.

(iii) Normal oxygen concentration is not inhibitory for the growth in contrast to C_3 plants.

Glycolic Acid or C_2 Pathway

Calvin et.al. (1950) reported that glycolic acid in labelled early during photosynthesis with $C^{14}O_2$. When the carbon dioxide concentration of the air is lowered below 1% glycolic acid is formed in large amount. Glycolic acid may be formed from ribulose-5-phosphate and, therefore, may be a secondary product of the photosynthetic carbon-reduction cycle, but, on the other hand, it may be direct product of photosynthesis. At high concentrations of carbon dioxide in the atmosphere, the synthesis of glycolic acid is inhibited and the phosphoglyceric acid is the favoured product. The mechanism of glycolic acid pathway is unknown. According to *Gibbs* (1971), biosynthesis of glycolic acid during photosynthesis coupling of photosynthetic carbon reduction cycle to the light reactions may be associated either with PS-I or PS-II.

Measurement of photosynthesis

Several methods have been used to measure the rate of photosynthesis. In one important and relatively accurate method, a single leaf, still attached to the plants, is enclosed in a small chamber and placed in the light. A continuous stream of air is then passed through the chamber, and the amount of carbon dioxide in the outgoing air is measured. This is compared with the amount of carbon dioxide in an identical stream of normal air. A reduction in the amount of carbon dioxide leaving the chamber as compared to the amount entering does not, however, indicate the real, or true, rate of photosynthesis but is termed the *apparent* photosynthesis.

It is not the true rate of photosynthesis because the living tissues of the leaf are, at the same time, oxidizing food in the process of respiration. In this energy-yielding process, oxygen is utilized and carbon dioxide released. Since this carbon dioxide, as well as that derived directly from the atmosphere, is used in photosynthesis, the amount of carbon dioxide released in respiration must also be determined and added to the apparent rate to determine the true rate. The true rate is therefore higher than the apparent rate. Another method of studying the rate of photosynthesis involves placing an aquatic plant in a jar of water in the sunlight. Bubbles of gas soon arise from the cut end of the stem.

If the plant is covered with a glass funnel and an inverted test tube filled with water placed over it, the gas will be collected as it displaces the water in the test tube. When a glowing wood splint is thrust into the test tube, it bursts into flame, indicating a high concentration of oxygen. The rate at which the bubbles arise roughly indicates the rate of photosynthesis. This, however, is subject to error, for the gas is not necessarily pure oxygen but may also contain nitrogen derived from the intercellular spaces of the leaf and stem. Moreover, some of the oxygen released in photosynthesis diffuses into the surrounding water and thus will not appear in bubble form. The rate at which organic matter in a plant increases during a given period of time may also be used as an index of the rate of photosynthesis. Several methods are used. In one of these, disks with an area of about 1 square centimeter are removed from one side of a leaf in the morning.

The *dry weight* of these disks is then obtained by placing them in an oven at 100°C and heating them until the weight no longer decreases. This takes about 24 hours. At the end of the day an equal number of disks is removed from the other half of the leaf and their dry weight determined. The difference in average dry weight between the disks removed in the morning and those

Distinction between green plant photosynthesis and bacterial photosynthesis.

Green plant photosynthesis	*Bacterial photosynthesis*
1. *Photosystems*: Consists of two photosystems (PSI and PSII) connected in series.	1. Consists of two photosystems connected parallely.
2. *Electron donor in PSII:* Water is the only electron donor for photoreduction.	2. H_2S, malate and succinate are the usual electron donors in photoreduction.
3. *Gas evolution* : Oxygen is evolved.	3. Oxygen is not evolved.
4. *Reductant* : The reductant of CO_2 is NADPH.	4. The reductant of CO_2 is NADPH.
5. *Electron transport chain*:	5.
(a) Cytochrome *b, c* present	(a) Cytochrome *b* is lacking.
(b) Plastocyanin is all the time present.	(b) Plastocyanin is absent.
(c) P_{700} and P_{690} are the reaction centres of PSI and PSII respectively	(c) One of the reaction centres is P_{890}, the other if any, is not known.
6. *Wavelength of light at which it occurs* : 400 -700 nm.	6. Above 700 nm.
7. *Emerson enhancement effect:* Detected.	7. Not detected.
8. *Special features* : Can be converted into bacteria type photosynthesis, of the plants are allowed to photo-synthesize in an atmosphere of hydrogen.	8. Can not be converted into green plant's type photosynthesis.

removed in the late afternoon is taken to represent the carbohydrate manufactured in an area the size of the sample during the day. This figure, however, is also subject to error, because (1) some of the sugar was respired during the day- that is, a gain in weight is a measure only of apparent photosynthesis; (2) some sugar and other foods were moved to other parts of the plant; and (3) the

Distinction between cytochrome and phytochrome

Cytochrome	*Phytochrome*
1. It is one of the electron carriers of mitochondrial membrane system, associated with phosphorylation.	1. It is a morphogenetic pigment found in all growing cells.
2. It is an iron containing cyclic tetrapyrrolic com pound. It has different forms - *Cyt b, Cyt c, Cyt a, Cyt* a_2 *etc.,*	2. It is a linear tetrapyrrole compound associated with protein. It has two absorbing forms-*phytochrome* red form (P_r) and *phytochrome* red from (Pfr). Both forms are inter-convertible. Red Light $P_r \longleftrightarrow P_{f7}$ Far Red Light of Darkenss
3. It carries 2e- at a time as noticed in both photophosphorylations (PSII) and oxidative phosphorylation. It is also associated with salt respiration, nitrate reduction etc.	3. It performs a number of physiological function such as (a) morphogenesis and differentiation (b) flowering (c) germination of seeds etc.,

area of the leaf may shrink during the day as a result of loss of water.

The "twin-leaf" method, a variation of the above, uses plants in which the leaves are opposite, arranged in pairs. A number of pairs are selected. The area of one leaf of each pair is determined. One member of each pair is removed early in the morning and its dry weight determined. This is compared with the dry weight of opposite leaves removed in the late afternoon. Any gain in dry weight is clearly evident. In spite of the fact that some of the sugar of the plant is lost in respiration, more than 90 percent of the dry weight of the plant results, directly or indirectly, from the photosynthetic activity of the leaves.

In terms of leaf area, the gain in dry weight may vary from 0.5 to 2 grams per square meter per hour. The magnitude of the photosynthetic process may better be visualized by the calculation that an acre of corn plants, producing 100 bushels, will manufacture on the average 200 pounds of sugar a day. Only about one third of this is deposited in the grain and thus becomes directly available as food for man. Even so, the products of photosynthesis are prodigious and it is obvious that investigations on the rate of this process are of practical importance.

Distinction between photosynthesis and respiration

Photosynthesis	*Respiration*
1. *Cellular site:* Chloroplast	1. Mitochondria
2. *Substrates:* CO_2 and water are converted to hexose sugar glucose ($C_6H_{12}O_6$) and other carbohydrates in the presence of light.	2. Glucose ($C_6H_{12}O_6$) is broken down into CO_2 and H_2O and energy is released. It take place mostly in the presence of oxygen.
3. *Products:* Glucose and other polysaccharides like starch, cellulose etc.,	3. CO_2, H_2O and energy in aerobic respiration; alcohol and lactic acid etc. in anaerobic or fermentative reactions.
4. *Nature of reaction:*Reductive process.	4. Oxidative process.
5. *Change of energy* : Kinetic energy (solar radiation) is converted to potential energy.	5. Potential energy is released in the form of kinetic energy (i.e., heat or ATP).

The Products of Photosynthesis

The products of photosynthesis are generally considered to be the carbohydrates and oxygen. Most of the latter diffuses out of the cells in which it is set free and plays no further part in the metabolism of the plant. Some of the oxygen, however,may be chemically intercepted and utilized in respiration within the cells. The products of importance in the plant nutrition are the energy-rich carbohydrates which are built up in the chloroplasts. Only four carbohydrates commonly appear in leaf cells during or immediately after photosynthesis. These are the hexoses. D-glucose and D-fructose, the disaccharide sucrose,and the polysaccharide starch. Numerous attempts have been products of photosynthesis

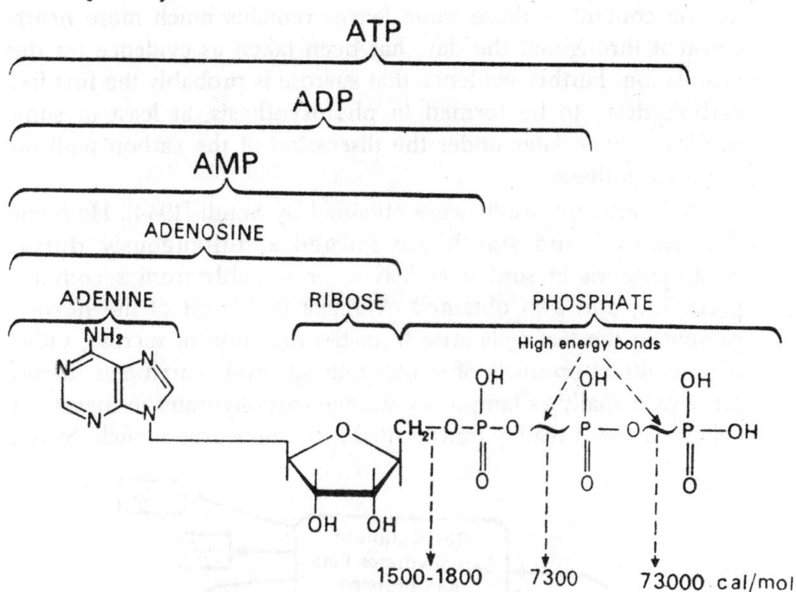

Fig. 2.16. ATP molecule-showing two type of energy bonds–last two phosphate bonds (wavy lines) release 7300 calories of energy. The first bond (straight line) release 1500-1800 calories of energy (AMP–Adenosine Monophosphate, ADP–Adenosine Di Phosphate, ATP–Adenosine Tri Phosphate).

and also which is the "first" product in this process,but to date there is no entirely satisfactory answer to this question. It appears quite probable, in fact, that the products of photosynthesis may differ somewhat from one photosynthetic tissue to another.

The work of some investigators seems to indicate that the principal leaf carbohydrates are formed sequentially in the order of their complexity. Weevers (1924), for example, showed that if geranium plants, the leaves of which had been depleted of sugars by keeping them in the dark, were allowed to photosynthesize for a very short period, only hexoses accumulated. A longer exposure was required for the appearance of sucrose, and a still longer one for the appearance of starch. On the other hand, a number of investigators have claimed that sucrose ($C_{12}H_{22}O_{11}$) is the first sugar to be formed in photosynthesis. The fact that the sucrose content to be formed in photosynthesis.

The fact that the sucrose content of the leaves of a number of species increases during periods of active photosynthesis and decreases rapidly upon the cessation of photosynthesis, while the hexose content of those same leaves remains much more nearly constant throughout the day, has been taken as evidence for this postulation. Further evidence that sucrose is probably the first free carbohydrate to be formed in photosynthesis, at least in some species,is given later under the discussion of the carbon pathway in photosynthesis.

Still different results were obtained by Smith (1944). He found that sucrose and starch are formed simultaneously during photosynthesis in sunflower leaves, presumably from a common precursor, and also obtained evidence that most of the hexoses present in the leaf cells arise from the digestion of sucrose, rather than as direct products of photosynthesis. Both starch and sucrose serve principally as temporary storage carbohydrates in mesophyll cells. Sucrose is readily translocated from the leaves as such. Starch,

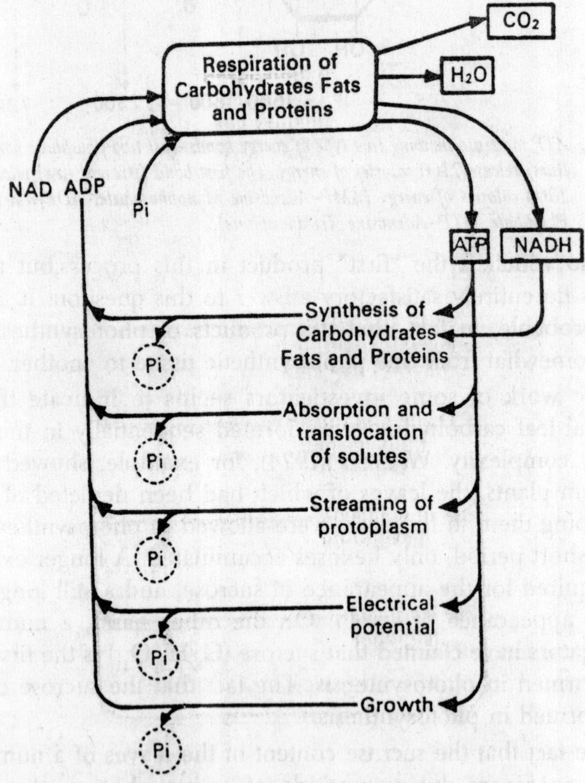

Fig. 2.17. Functions of ATP.

however, being insoluble, is immobile until digested to simpler, soluble carbohydrates which can move out of the mesophyll cells. There is some evidence that other types of stable organic compounds may arise more or less directly in the photosynthetic process, at least undersome conditions or in some species.

Burstrom's (1943) results with young, excised wheat leaves indicate that a direct reduction of nitrates resulting in the formation of amino acids or related compounds may occur as an integral part of photosynthesis. Indications of a similar close relationship between nitrogen metabolism and photosynthesis has also been found in *Chlorella.* Evidence of an intermeshing of the processes or photosynthesis and amino acid synthesis also comes from studies of the carbon pathway in photosynthesis as revealed by the use of radioactive isotopes. Results indicating that amino acid synthesis occurs in young leaves of at least some species as an apparently integral part of photosynthesis contrasts with the results of Smith (1943), who found an exact equivalence between the quantity of carbon dioxide photosynthesized and the quantity of carbohydrates made in sunflower leaves. It seems highly probable that the course of photosynthesis is not exactly the same in all photosynthetic tissues. The process may follow a different course in young, growing leaves, for example, than it does in older and more mature leaves, even in the same species.

Starch synthesis

One of the most familiar facts about photosynthesis is that starch accumulates in the chloroplast of many species shortly, if not immediately, after the onset of the process at any appreciable rate. Although the possibility of the direct formation of starch in photosynthesis in some species cannot be excluded, in general starch synthesis and photosynthesis are independent processes. Since, starch is insoluble in water the presence of starch grains in a cell must signify that the grains have been built up in that cell. Starch grains often occur abundantly in the cells of non-green tissues or in the tissues of roots or other plant organs which are never exposed to light.

Obviously photosynthesis cannot take place in such cells and the starch present must have been synthesized from sugars translocated to them from the green organs of the plant. Since starch synthesis in the non-green cells of plants is obviously a distinct process from photosynthesis, it seems probable that this is

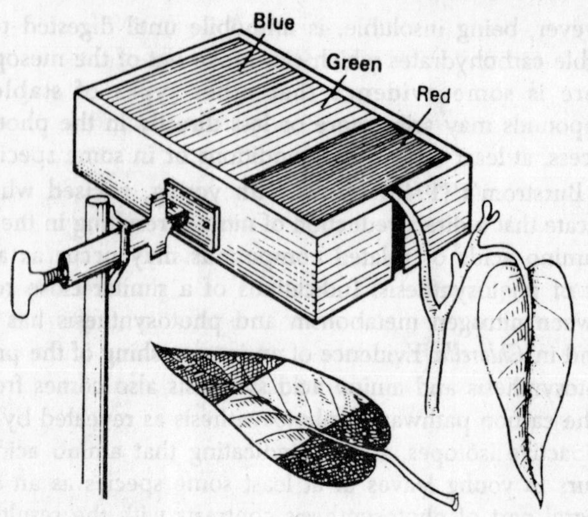

Fig. 2.18. Ganong's light screen to study the effect of light of different colours on the photosynthesis.

also true of the starch synthesis which occurs in green cells. The fact that starch synthesis will take place in the chlorolasts of green leaves floated on a sugar solution in the dark also supports this view. Photosynthesis occurs only in the present of light and in the higher plants, only in the chloroplasts. Starch synthesis may occur in the chloroplasts, most probably as the product of a secondary reaction, but also may take place in many non-green cells and in either the light or dark. In non-green cells starch is synthesized in the leucoplasts. The critical concentration of simple sugars required for starch formation in the leaves of many species is very low and has been reported to be less than 0.5g per 100g. of fresh leaves in some species.

In the mesophyll cells of most species,therefore, starch formation in the leaves quickly follows photosynthesis, much of the sugar made in the latter process being converted into starch. Under conditions favourable to photosynthesis the starch content of the leaves of most species usually increases during much of the daylight period. During the night hours the starch content of leaves usually decreases because of digestion out of the cells in the form of this sugar or other soluble carbohydrates produced there from. That starch synthesis is an entirely independent process from photosynthesis is also indicated by the fact that this process does not occur in the mesophyll cells of a number of species of plants, yet photosynthesis takes place in these cells in the same manner as in all other green plants. Failure of the leaves to synthesize starch is a characteristic feature of the metabolism of many species of the Liliaceae, Amaryllidaceae, gentianaceae, Compositae, and Umbelliferae. Similarly the non-green portions of at least some kinds of variegated leaves do not normally synthesize starch. However, if the sugar concentration within the chloropohyll-free cells is artificially increased by floating such leaves on glucose solutions, starch synthesis can be induced. For the leaves of the variegated gernium a glucose solution of about 0.5M concentration has been found to be optimum for the induction of starch synthesis in the non -green portions.

The measurement of photosynthesis

Measurements of photosynthesis are complicated by the fact that certain other processes involving the same materials are proceeding in the cells at the same time. The process of respiration is continually in progress in all cells, resulting in an oxidation of part of the carbohydrates synthesized in photosynthesis. This introduces an error which is inherent in all methods of measuring photoysnthesis. Determination of the quantity of photosynthate formed it a given time are always less than the true value by the amount of carbohydrate which has been consumed in respiration. In many measurements of photosynthesis the simultaneous occurence of respiration is disregarded, and the results obtained are designated at the *apparent* or net photosynthetic rate, or in other words, as the rate of photosynthesis minus the rate of respiration.

Since in rapidly photosynthesizing tissues the rate of photosynthesis if often ten to twenty times as great as the rate of respiration, the apparent photosynthetic rate is often not greatly less than the true rate. For many purposes measurements of the rate of apparent photosynthesis have as great or greater significance as determinations of the "true" rate. Values for the actual rate of photosynthesis are obtained by correcting apparent rates for the quantity of carbohydrates consumed in respiration during the period of the measurement. The values used for such corrections are obtained by measuring the respiration rate of the same plant or organ when treated so that photosynthesis cannot occur–as, for example, by subjection to complete darkness. Quantitative measurements of photosynthesis in terrestrial plants are usually based on determinations of the rate of oxygen evolution or rate of carbon dioxide consumption. The plants or parts of plants are enclosed in transparent chambers or leaf cups are attached to the leaves. Outside air is passed through the chambers or leaf cups at a relatively rapid rate. The air is collected upon emergence from the chamber and analyzed for oxygen, or carbon dioxide, or both. Another procedure is to analyze small samples of the air for these gases at frequent intervals.

By a comparison of the results of these analyses with similar analyses of the outside atmosphere, the rate of carbon dioxide consumption or oxygen evolution can be computed, either of which will serve as an index of the rate of apparent photosynthesis. The rate of carbon dioxide consumption, under field conditions favourable for photosynthesis, is usually between 10 and 20 mg. CO_2 per cm^2 or leaf area per hr. Many of the most fundamental investigations of photosynthetic rates have been made by very sensitive manometric techniques which are intrinsically limited in their application to determinations of photosynthesis in small units of plant tissue.

Manometric methods of measuring photosynthesis have been used most widely with suspensions of algae such as *Chlorella.* In these techniques the rate of oxygen evolution, or carbon dioxide consumption, or both, are measured by pressure changes in a closed system in which the photosynthesizing tissue is enclosed. The simplest method of making rough quantitative measurements of the rate of photosynthesis interrestrial plants is to determine the increase in the dry weight of leaves during a given interval of time. One method of procedure is as follows : a representative sample of disks, usually 100 or more, is cut from the leaves by means of a cork borer at the beginning of the period for which the determination is to be made. These disks are transferred to an oven and dried to constant weight. At the end of a chosen period of time, the plant having been meanwhile exposed to the desired environmental conditions, a second sample consisting of the same number of disks as the first is cut from the leaves.

Various precautions must be taken to insure the two samples being as nearly comparable as possible. The dry weight of th second sample of disks is also determined. Any gain in dry weight is considered to represent carbohydrates which have accumulated in the leaves as a result of photosynthesis. Ths method is subject to several serious errors and yields only approximate results. The gain in dry weight of leaves under conditions favourable for photosynthesis is usually between o 0.5 and 2.0g. per square meter per hour. The "twin leaf" method described by Denny (1930) is a variation of this method.

The use of isotopes as tracers in physiological investigations

The atoms of most chemical elements exist in more than one variety. Each kind of a given element has a different atomic weight, but all of them carry the same nuclear charge. For example,there are three different kinds of magnesium atoms with atomic weights of 24, 25, and 26, respectively. Such different varieties of atoms of a given element are called *isotopes.* Differences in the chemical behaviour of two isotopes of the same element are sc slight as to be barely detectable, and ordinarily they cannot be separated by chemical methods.

The chemical properties of all isotopes of a given element are virtually identical, because all have the same electron configuration; One isotope differs from another only in the constitution of the atomic nucleus which is composed of protons and neutrons. In addition to the stable isotopes described above there are many radioactive isotopes of various elements. Radioactive istopes of certain heavy elements such as uranium and radium occur in nature. The atoms of many elements not normally radioactive may be made so by bombardment with various types of elementary particles such as neutrons, protons, deutrons, and alpha particles. Such bombardments are usually accomplished with the aid of a cyclotron or an uranium pile reactor. Atoms of all radioactive elements gradually disintegrate, emitting charged particles and radiations in the process.

For the artificially radioactive elements which are of greatest significance in biological sciences, the charged particles emitted are electrons and positrons. As a result of the loss of particles and radiations by radioactive isotopes one element is transmitted into another. Radioactive Na^{24}, for example, with the loss of an electron, is converted into the stable naturally occurring Mg^{24} isotope. The *half-life* of a radioactive isotope is the length of time required for half of any given weight of the isotope to be converted into the kind of atom into which it is degraded. The half-lives of radioactive isotopes range from less than a second to many hundreds of years. The situation with regard to isotopes is well illustrated by the biologically important element carbon. Naturally occuring carbon consists almost entirely of the stable C^{12} isotope. The C^{13} isotope is also stable and occurs intraces in nature.

The atomic weight of naturally occurring carbon is a little more than 12, because of the presence of a slight admixture of the C^{13} isotope. In addition, three radioactive isotopes can be made: C^{10}, with a half-life of 9 sec, C^{11}, with a half -life of 21 min: and C^{14}, with a half-life of about 5000 years. A radioactive isotope betrays its presence wherever it may be by the continual emission of charged particles which can be detected with an electroscope of Geiger counter.

Similarly, the presence of any of rarer stable isotopes in a system can be detected, because of their unusual mass, by means of the instrument known as the mass spectograph. Hence, if molecules of any compound are tagged, by incorporating into them atoms that are radioactive or atoms of the rarer kinds of stable istopes, it is possible to trace such molecules through a system. After the absorption or ingestion of such molecules by an organism it is possible to follow their subsequent distribution through the organism and often to determine the reactions in which they participate. This is not possible by ordinary methods of chemical analysis because by such methods it is impossible to distinguish between introduced molecules and others of the same species which were already present in the organism.

3

CARBON IN PHOTOSYNTHESIS

The processes of life consist ultimately of the synthesis and breakdown of carbon compounds. Because a carbon atom can bind four other atoms to itself at a time and is thereby able to link up with other atoms-especially other carbon atoms in chains and rings, carbon lends itself to the construction of a virtually endless variety of molecules. These molecules derive their physical characteristics and chemical activity not only from their composition but also from their size and intricacy of structure. The rich variety of life suggests in turn that living cells have gone far in the elaboration of such composition and the processes that make and unmake them.

All these processes depend in the end on a first one. This is the process of photosynthesis, which takes carbon and several other common elements from the environment and builds them into the substances of life. The plant finds most of these elements already bonded to oxygen in oxides such as carbon dioxide (CO_2), water (H_2O), nitrate (NO_2^-) and sulfate (SO_4^{--}). Before the plant can bind the elements other than oxygen together as organic compounds, it must remove some of the excess oxygen as oxygen gas (O_2), and this accomplishment takes a large amount of energy.

In the simplest terms photosynthesis is the process by which green plants trap the energy of sunlight by using that energy to break strong bonds between oxygen and other elements, while forming weaker bonds between the other elements and forcing oxygen atoms to pair as oxygen gas. For example, to make the sugar glucose ($C_6H_{12}O_6$) the plant must split out six molecules of oxygen in order to combine the carbon and half the oxygen of six

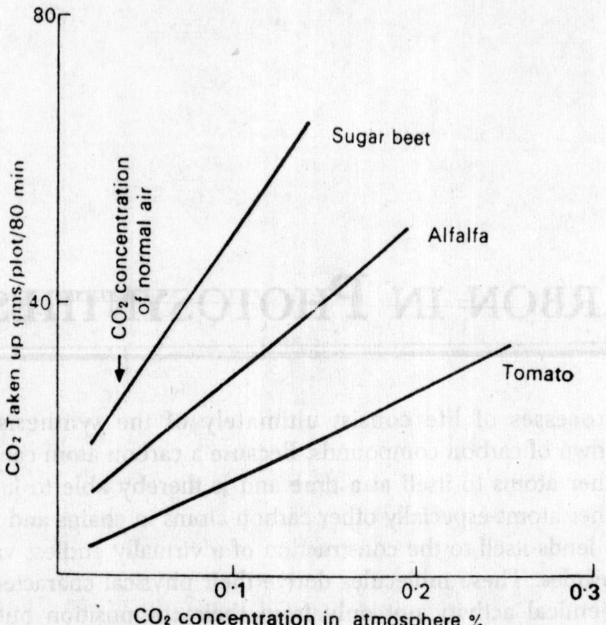

Fig. 3.1. Effect of CO_2 concentration on the rate of photosynthesis.

carbon dioxide molecules with the hydrogen of six water molecules. The glucose and other organic compounds taken up in the chemical machinery of the plants and the animals that live on plants serve both as fuel and as the raw materials for the synthesis of higher organic compounds. That considerable solar energy is bound by photosynthesis becomes apparent when wood or coal is burned.

In living cells the controlled combustion of respiration extracts this energy to power the other processes of life. Both kinds of combustion take oxygen from the air and break down organic compounds to carbon dioxide and water again. In its end result photosynthesis can be defence as the opposite of respiration. Together these complementary processes drive the cyclic flow of matter and the noncyclic flow of energy through the living world. From such generalizations about the effect and function of photosynthesis in nature it is a long step to the explanation of how photosynthesis works. Yet much of the explanation is now complete.

The work has been greatly facilitated by the earlier and more nearly complete resolution of the chemistry of respiration. The two processes, it turns out, are in some ways comple-mentary on the molecular scale, just as they are on the grand scale in the biosphere. Each involves some 20 to 30 discrete reactions and as many intermediate compounds; half a dozen of these reactions and their intermediates are common to both photosynthesis and respiration. Only the first few stops in photosynthesis are driven directly be light.

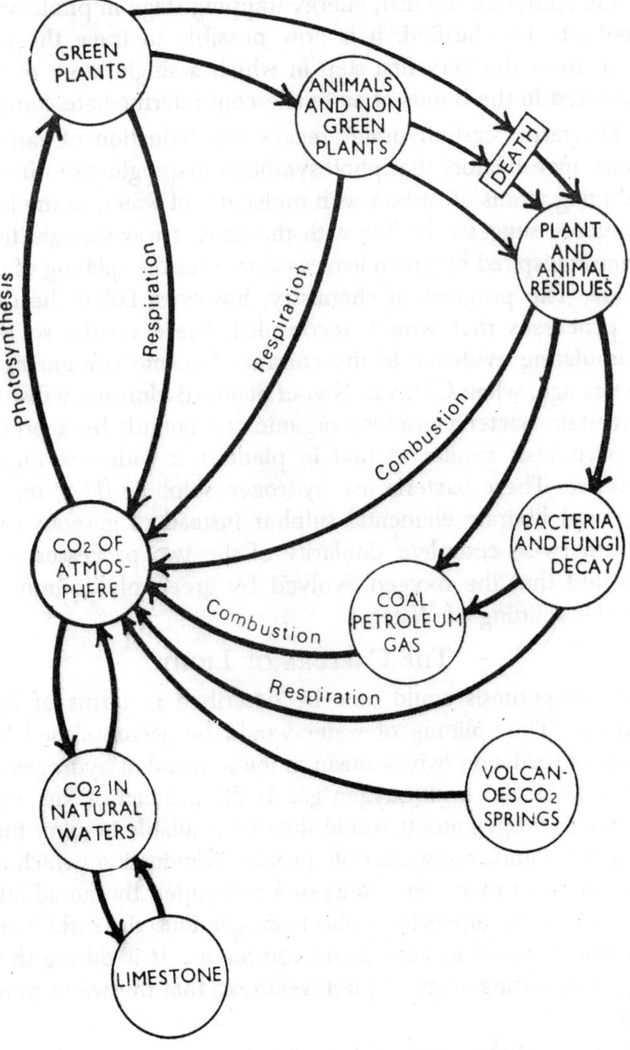

Fig. 3.2. Carbon dioxide cycle in nature.

The energy of light is trapped in the bonds of a few specific compounds. These energy carriers deliver the energy in discrete units to the steps of synthesis that follow. The same or closely similar carriers perform corresponding operations in respiration, picking up energy from the stepwise dismemberment of the fuel molecule and delivering it to the energy consuming processes of the cell. Although the first, energy-trapping stage in photosynthesis remains to be clarified, it is now possible to trace the path of carbon from the very first step in which a single atom of carbon is captured in the bonds of an evanescent intermediate compound.

The term "carbohydrate" recalls the deduction of early 19th-century investigators that photosynthesis made glucose directly by combining atoms of carbon with molecules of water, as the formula for glucose suggests. In line with this idea it was thought that the oxygen transpired by green leaves came from the splitting of carbon dioxide. The progress of chemistry, however, failed the disclose any processes that would accomplish these results so simply. Accumulating evidence to the contrary became convincing same 60 years ago, when C.B. van Niel of Stanford University discovered that certain bacteria produce organic compounds by a process of photosynthesis similar to that in plants but with one important difference. These bacteria use hydrogen sulphide (H_2S) instead of water and liberate elemental sulphur instead of gaseous oxygen. The otherwise complete similarity of the two processes strongly suggested that the oxygen evolved by green plants must come from the splitting of Water.

The Capture of Light

Photosynthesis could now be described in terms of familiar chemistry. The splitting of water would be accomplished by the process of oxidation (which mean as the removal of hydrogen atoms from a molecule), with oxygen gas as the product of the reaction. The free hydrogen atoms would then be available to carry through the equally familiar and opposite process of reduction (which means the addition of hydrogen atoms to a molecule). By the addition of hydrogen atoms (or electrons plus hydrogen ions) the carbon dioxide would be reduced to an organic compound. It is during the first, energy-converting stage of photosynthesis that the water molecule is split.

Initially the energy of light impinging on the plant cell is transformed into the chemical potential energy of electrons excited

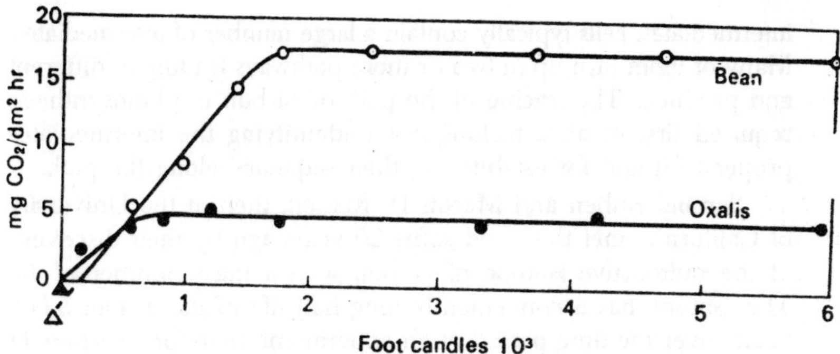

Fig. 3.3. Effect of intensity of light on the rate of photosynthesis in bean and Oxalis.

from their normal orbits in molecules of the green pigment chlorophyll and other plant pigments. A large part of this energy eventually goes into the splitting of water as electrons and hydrogen ions are transferred from water to the substance triphosphopyridine nucleotide (TPN^+), which is thereupon reduced to the form designated TPNH. The TPNH thus becomes not only a carrier of energy but also the bearer of electrons for the subsequent reduction of carbon dioxide. Along with the movement of electrons from water to TPNH, some energy goes to charging the energy-carrying molecule adenosine triphosphate (ATP), specifically by promoting the attachment of a third phosphate group ($-OPO_3^{--}$) to adenosine diphosphate (ADP), the discharged form of the carrier.

Both ATP and TPNH belong to the family of compounds known as cofactors or coenzymes, which work with enzymes in the catalysis of chemical reactions. ATP, the universal currency of energy transactions in the cell, plays a significant role in respiration as well as in photosynthesis. Needless to say, the manufacture of each of these cofactors involves an intricate cycle of reactions. Although the cycles are not yet fully understood it is enough for the purpose of the present discussion to know that ATP and TPNH, or closely similar compounds, furnish the energy for the second stage of photosynthesis, during which the carbon atom of carbon dioxide is reduced and joined to a hydrogen atom and a carbon atom in place of an oxygen.

The process of reduction goes forward in small steps. Each reaction brings about some change in a carbon compound until the starting material is at last transformed to the final product. For each reaction there is therefore an intermediate compound. Since every life process involves a more or less extended series of intermediates, cells typically contain a large number of intermediates. Many of them turn up in two or more pathways leading to different end products. The tracing of the path of carbon in photosynthesis required first of all a technique for identifying the intermediates proper to it and for establishing their sequence along the path.

Samuel Ruben and Martin D. Kamen, then at the University of California, met this need some 20 years ago by their discovery of the radioactive isotope of carbon with a mass number of 14. This isotope has a conveniently long half life of more than 5,000 years; over the time period of an experiment, therefore, carbon 14 has an effectively constant radioactivity. Ruben and his colleagues recognized at once the potential usefulness of this isotope as a label for the identification of compounds in biological processes. They prepared carbon dioxide in which the carbon atoms were carbon 14. When they exposed green plants to an atmosphere containing this gas instead of normal carbon dioxide ($C^{12}O_2$), the plants look up the $C^{14}O_2$ and made compounds could be detected by various devices, such as the Ceiger-Muller counter, and by radioautography on X-ray film. Unfortunately this work was cut short by the war and by Ruben's death in a laboratory accident.

In 1946 Melvin Calvin organized a new group at the Lawrence Radiation Laboratory of the University of California with the primary objective of tracing the path of carbon in photosynthesis with $C^{14}O_2$ as one of its principal tools. The early experiments were quite simply contrived. We used leafy plants and often just the leaves of plants. After allowing a leaf to photosynthesize for a given length of time in an atmosphere of $C^{14}O_2$ in a closed chamber, we would bring biochemical activity to a halt by immensing the leaf in alcohol. With the enzymes inactivated, the reactions converting one intermediate compound into another would stop, and the pattern of labelling would be "frozen" at that point. We soon discovered, however, that photosynthesis proceeds too rapidly for completely reliable observation by such a procedure.

With a few seconds delay in the penetration of alcohol into the cell, for example, the labelling pattern would be disarrayed and no longer representative of the stage at which we tried to halt the photosynthesis. Since rapid and precisely timed killing of the plant is important, we adopted single-celled algae–*Chlorella pyrenoidosa* and *Scenedesmus obliquus*–as the subject for many of our experiments. In both species the plant consists of a single cell so small that it can be seen only with a microscope. Alcohol can quickly penetrate the cell wall and deactivate the enzymes. The algae offer another advantage: they can be grown in continuous cultures, assuring us a supply of material with highly constant properties. An experimental sample is taken from the culture in a thin walled, transparent closed vessel. Illuminated through the walls of the vessel and supplied with a stream of ordinary carbon dioxide, which is bubbled through the suspension, the algae photosynthesize at the normal rate. The supply of carbon dioxide was shut off and a solution of radioactive bicarbonate ion (carbon dioxide dissolved in algae culture medium is mostly converted into bicarbonate ion) was injected. After a few seconds or minutes the cells are killed. We then extract the soluble radioactive compounds from the plant material and analyze them.

The Reduction of CO_2

Calvin and his colleagues soon found that the carbon 14 label was distributed among several classes of biochemical compounds, including not only sugars but also amino acids: the subunits of proteins. As the exposure time was reduced to a few seconds, the first stable intermediate product of photosynthesis was found to be the three-carbon compound 3-phosphoglyceric acid (PGA). The next step was to determine which of the three carbon atoms in the first generation of PGA molecules synthesized in the presence of radioactive carbon dioxide bears the carbon 14 lable. We first removed from PGA the phosphate group and then diluted the free glyceric acid with glyceric acid containing the stable carbon 12 isotope in order to have enough material for analysis by ordinary chemical methods. Treatment with reagents that severed the bonds between the carbons produced three different products, one from each carbon atom. By measuring the radioactivity of each of the products we were able to identify the labelled carbon.

In PGA from plants that had been exposed to labelled carbon dioxide for only five seconds we found that virtually all the carbon 14 was located in the carboxyl atom, the carbon at one end of the chain that is bound to two oxygens. This was not surprising because the carboxyl group most nearly resembles carbon dioxide. The carbon is bound to the oxygen by three bonds, however, instead of four; the fourth bond now ties it to the middle carbon in the PGA chain. The transfer of this bond from one of the oxygen to a carbon constitutes the first step in the reduction of the carbon

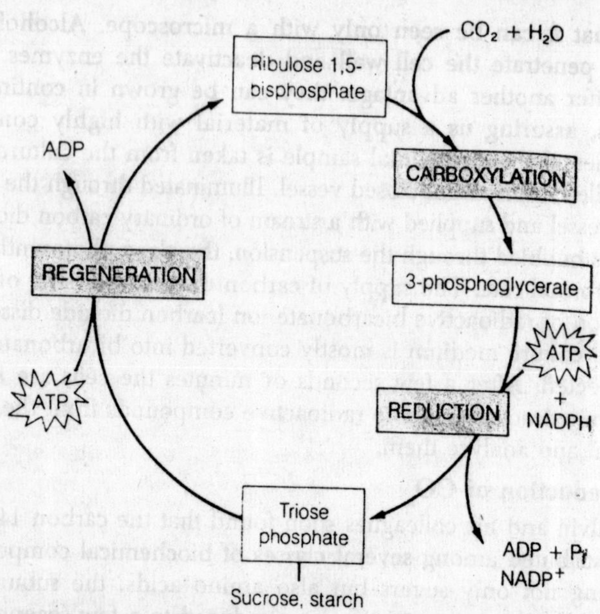

Fig. 3.4. The C3 photosynthetic carbon reduction cycle (PCR cycle) proceeds in three stages: (1) carboxylation, during which CO_2 is covalently linked to a carbon skeleton; (2) reduction, during which carbohydrate is formed at the expense of the photochemically derived ATP and reducing equivalents, NADPH; and (3) regeneration, during which the CO_2-acceptor molecule, ribulose 1,5-bisphosphate is re-formed.

dioxide. This was evidence also that the reduction is accomplished by some sort of carboxylation reaction, a reaction, in which carbon dioxide is added to some organic compound. Ultimately, of course, the two other carbons of PGA must come from carbon dioxide. But it was some time before investigation disclosed the specific compound that picks up the carbon dioxide and the cyclic pathway that makes this carbon dioxide acceptor from PGA. The discovery of the pathway intermediates was made much easier by a then comparatively new technique : two-dimensional paper chromatography, developed by the British chemists A.J.P. Martin and R.L.M. Synge.

Closely similar compounds can be distinguished in this procedure by slight differences in their relative solubility in an organic solvent and in water. The extract from the plant is dropped on a sheet of filter paper near one corner. An edge of the paper adjacent to the corner is immersed in a trough containing an organic solvent; the paper is held take by weight and the whole assembly is placed in a water-saturated atmosphere in a vapour tight box. The solvent travelling through the paper by capillarity dissolves the compounds and carries them along with it. As they move along in the solvent, however, the compounds tend to distribute themselves between the solvent and the water absorbed by the fibers of the paper. In general the more soluble the compound is in water compared with the organic solvent, the slower it travels. If the compound is also absorbed to some extent by the cellulose fibers, its movement will be even slower. As a result the compounds are distributed in a row in one dimension. Depending on the solubility of the compounds and the nature of the solvent used, some compounds may still overlap one another. Repetition of the procedure, with a different solvent travelling at right angles to the direction of the first run, will usually separate these compounds in a second dimension.

Since most of the compounds are colourless, special techniques are needed to locate them on the paper. Those that are radioactive will locate themselves, however, if the chromatographic paper is placed in contact with a sheet of X-ray film for a few days. The resulting radioautograph will show as many as 20 or 30 radioactive compounds in the substances extracted from algae exposed to carbon 14 for only 30 seconds. Clearly the synthetic apparatus of the plant works rapidly.

Chromatographs and Radioautographs

In order to identify these compounds, a chromatographic map was prepared by running samples of many known compounds through the same chromatographic system and recording the locations at which they were found on the paper. The locations can be made visible in these cases by spraying the paper with a mist of some chemical that is known to react with the compound of produce a coloured spot. Comparison of the radioautograph of an unknown compound with the map yields a first clue to its identity. This can be corroborated by washing the radioactive compound out of the paper with water and mixing it with a larger sample of the suspected authentic substance. The mixture is applied to a new piece of filter paper and chromatographed. with enough of the authentic material to yield a coloured spot, comparison with a radioautograph of the same paper now shows whether or not the radioactive and authentic material really coincide. The possibility that the authentic material and the radioactive material are still not the same can be tested by using different solvent systems in the preparation of the chromatograph and by other means.

Over the years these procedures have established the identity of a great many of the intermediate and end products of photosynthesis. Some of the sugar phosphates labelled by carbon 14 proved to be well-known derivatives of triose (three carbon) and hexose (six-carbon) sugars. Others were discovered for the first time among the intermediates produced by our algae. Benson showed that among these are a seven-carbon sugar phosphate and also five-carbon phosphates, including in particular ribulose-1,5-diphosphate.

The rapid building of carbon 14 into the more familiar triose and hexose phosphates suggested certain biochemical pathways already established in studies of respiration. It seemed likely that PGA might be linked to these phosphates by the reverse of a sequence of respiratory reactions first mapped many years ago by the German chemists Otto Meyerhof, Gustav Embden and Jakob Parnas. In the respiratory pathway hexose phosphate is split into two molecules of triose phosphate, with the split occuring between the two carbon atoms in the middle of the chain. The triose phosphate is then oxidized to give PGA. The electrons from this energy yielding operation are picked up by diphosphopyridine nucleotide (DPN^+), which is thereupon reduced to DPNH. The DPN^+ is a close relative of the TPN^+ that turns up in photosynthesis. In addition this oxidation yields enough energy to make a molecule of ATP from ADP and phosphate ion.

In the reverse pathway of these reactions in photosynthesis. Calvin concluded, the plant uses the factors ATP and TPNH, made earlier by the transformation of the energy of light, to bring about the reduction of PGA to triose phosphate. In the first step the terminal phosphate group of ATP is transferred to the carboxyl group of PGA to form a "carboxyl phosphate" (really an acyl phosphate). Some of the chemical potential energy that was stored in ATP is now stored in the acyl phosphate, making the new intermediate compound highly reactive. It is now ready for reduction by TPNH. This reducing agent donates two electrons to the reactive intermediate. On carbon-oxygen bond is thereby severed and the oxygen atom, carried off with the phosphate group, is replaced by a hydrogen atom. In this way the carboxyl carbon atom is reduced to an aldehyde carbon atom; that is, it now has two bonds to oxygen instead of three and one bond each to carbon and hydrogen. This is the point in the cycle at which most of the solar energy captured in the first stage of photosynthesis is applied to the reduction of carbon.

The Unstable Intermediate

The next development in the plotting of the carbon pathway came from a series of experiments first performed by Peter Massini. He hoped to see which intermediates would be most strongly increased or decreased in concentration by turning off the light and allowing the synthetic process to go on for a while in the dark. In order to establish the concentration of the various intermediates when the reaction proceeds in the light, he bubbled radioactive carbon dioxide through the culture for more than half an hour. At the end of this period every intermediate was as highly radioactive as the incoming carbon dioxide. The radioactivity from each compound therefore gave a measure of the concentration of the compound. He then turned off the light and after a few seconds took another sample of algae in which he measured the relative concentration of compounds by the same technique. Comparison with the compounds sampled in the light showed that the concentration of PGA was greatly increased. This finding could be readily explained: turning off the light stopped the production of the ATP and TPNH required to reduce PGA to triose phosphate.

Of the sugar phosphates present, only one, the five-carbon ribulose diphosphate, was found to have changed significantly; its concentration dropped to zero. Because the PGA had simultaneously increased in concentration, it was apparent that ribulose diphosphate was consumed in the production of PGA. This finding was of great significance because it indicated for the first time that ribulose diphosphate is the intermediate to which carbon dioxide is attached by the carboxylation reaction.

For this reaction ribulose diphosphate is prepared by an earlier reaction that goes on in the light and in which ATP donates its terminal phosphate group to ribulose monophosphate. The more reactive diphosphate molecule now adds one molecule of carbon dioxide by carboxylation. The details of this reaction remain obscure because the resulting six-carbon intermediate is unstable and could not be detected by the methods of analysis. As its first stable product this sequence of events yields two three-carbon PGA molecules.

Massini's experimental results were confirmed by a parallel experiment devised by Alex Wilson. Instead of turning out the light Wilson shut off the supply of carbon dioxide. In this situation one might expect to find an increase in the concentration of the compound that is consumed in the carboxylation reaction; ribulose diphosphate showed such an increase. Correspondingly, one would look for a decrease in the concentration of the product of this reaction; PGA did in fact decrease in concentration.

The first steps along the path were thus established. The photosynthesizing plant starts with ribulose monophosphate and converts it to ribulose diphosphate, using chemical potential energy trapped from the light in the terminal phosphate bond of ATP. Carbon dioxide is joined to this compound, and the resulting six-carbon intermediate splits to two molecules of PGA. With energy and electrons supplied by ATP and TPNH, PGA is reduced to triose phosphate. In the next step, it was apparent, two triose phosphates must be joined end to end in the reverse of a familiar respiratory pathway to form a hexose phosphate. The pathway from hexose to pentose phosphate remained to be uncovered.

We continued the carbon-by-carbon dissection and analysis of these chains by the methods that had earlier shown the carbon 14 in PGA to be located first in the carboxyl carbon. In the hexose molecules we had found the labelled carbon concentrated in the two middle carbons, just where it should be if two triose molecules made from PGA were linked together by their labelled ends. We also took apart the seven-carbon and five-carbon sugar phosphates to establish the position of the carbon 14 atoms in their chains. As the result of these degradations we were able to show that the overall economy of the photosynthetic process starts with five three-carbon PGA's, variously transforms them through three-, six-, four- and seven-carbon phosphate intermediates and returns three five-carbon ribulose diphosphates to the starting points 3.2 and 3.5. From carboxylation of these three chains and their immediate bisection, the cycle at last yields six PGA molecules. The net result, therefore, is the conversion of three carbon dioxide molecules to one PGA molecule.

The Calvin Cycle

With these steps filled in, the carbon reduction cycle in photosynthesis, called the Calvin cycle, was complete. The intermediates formed in the cycle depart from it on various pathways

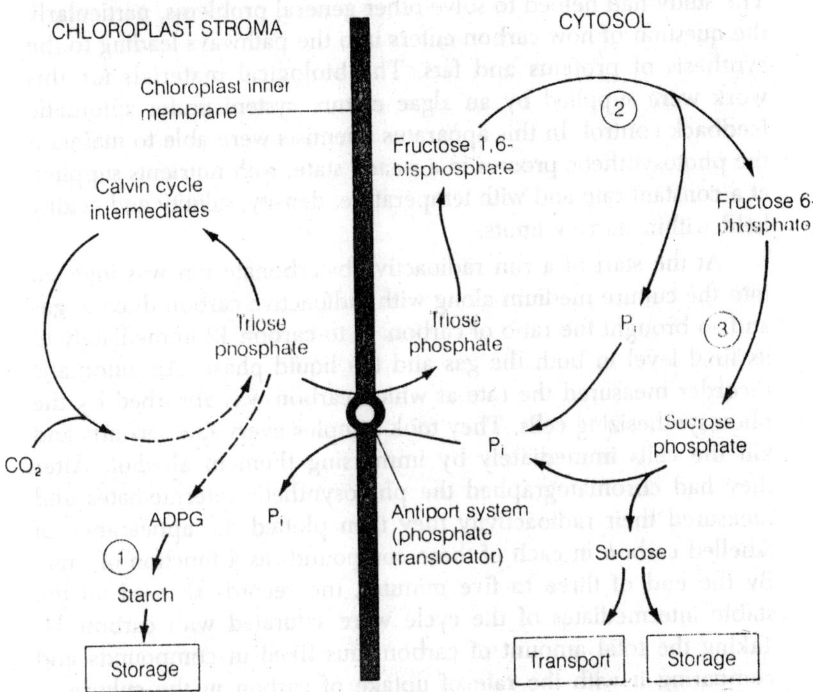

Fig. 3.5. A simplified scheme for starch and sucrose synthesis. Triose phosphate formed during photosynthesis in the Calvin (C3 photosynthetic carbon reduction) cycle, can either be utilized in starch formation in the chloroplast or transported into the cytosol in exchange for inorganic phosphate (P_i) via the phosphate translocator in the inner chloroplast membrane.

to be converted to the end products of photosynthesis. From triose phosphate, for example, one sequence of reactions leads to the six-carbon sugar glucose and the large family of carbohydrates. Because the cycle had been established primarily by experiments with algae and the leaves of a few higher plants, it was important to see whether or not the cycle prevailed throughout the plant kingdom. Calvin and Louisa and Richard Norris carried out experiments with a wide variety of photosynthetic organisms.

In every case, although they found variation in the amounts of particular intermediates formed, the pattern was qualitatively the same. It also had to be shown that the pathway we had trace out is quantitatively the most important route of carbon reduction in photosynthesis. To this end an intensive study of the kinetics of the flow of carbon in photosynthesis was carried out by scientists.

The study had helped to solve other general problems, particularly the question of how carbon enters into the pathways leading to the synthesis of proteins and fats. The biological materials for this work were supplied by an algae culture system under automatic feedback control. In this apparatus scientists were able to maintain the photosynthetic process in a steady state, with nutrients supplied at a constant rate and with temperature, density, salinity and acidity held within narrow limits.

At the start of a run radioactive bicarbonate ion was injected into the culture medium along with radioactive carbon dioxide gas and so brought the ratio of carbon 14 to carbon 12 immediately to its final level in both the gas and the liquid phase. An automatic recorder measured the rate at which carbon was absorbed by the photosynthesizing cells. They took samples every few seconds and kill the cells immediately by immersing them in alcohol. After they had chromatographed the photosynthetic intermediates and measured their radioactivity they then plotted the appearance of labelled carbon in each of these compounds as a function of time. By the end of three to five minutes, the records showed all the stable intermediates of the cycle were saturated with carbon 14. Taking the total amount of carbon thus fixed in compounds and comparing it with the rate of uptake of carbon in the culture, it was found that the cycle accounts for more than 70 percent of the total carbon fixed by the algae.

A small but significant amount was also taken up by the addition of carbon dioxide to a three-carbon compound, phosphoenolpyruvic acid, to give four-carbon compounds. From the earliest work with carbon 14 it had been apparent that carbon dioxide finds its way rather quickly into products other than carbohydrates in the photosynthesizing plant. This was at variance with traditional ideas about photosynthesis that regarded carbohydrates as the sole organic products of the process. It was important to ask, therefore, whether fats and amino acids could be formed directly from the cycle as products of this intermediates or whether these non-carbohydrates were synthesized only from the carbohydrate end products of photosynthesis.

The kinetic studies show that certain amino acids must indeed be formed from the intermediates and must therefore be regarded as true products of photosynthesis. The amino acid alanine, for example, shows up labelled by carbon 14 at least as rapidly as any carbohydrates; it would be labeled with carbon 14 much more slowly is it were made from carbohydrate, since the carbohydrate would have to be labeled first.

Scientists have been able to show that more than 30 per cent of the carbon taken up by the algae in our steady state system is incorporated directly into amino acids. There is some evidence that fats may also be formed as products of the cycle. The discovery that plants make these other compounds as direct products of photosynthesis lends new interest and importance to the chloroplast, the subcellular compartment of green cells that contains pigments and the rest of the photosynthetic apparatus. It has been known for some time that this highly structured organelle is responsible for the absorption of light, the splitting of water and the formation of the cofactors for carbon reduction. More recent studies have shown that it is the site of the entire carbon-reduction cycle. Now the chloroplast emerges as a complete photosynthetic factory for the production of just about everything necessary to the plant's growth and function.

4

Role of Chlorophyll in Photosynthesis

The Chloroplast Pigments

Green is the predominant colour of the plant kingdom. With only a few exceptions all leaves are green in colour as are also many other plant organs such as herbaceous and young woody stems, young fruits, and sepals of flowers. The green colouring of plants is often termed chlorophyll, although actually a number of kinds of chlorophyll occur in plants. Less evident is the fact that the leaves and many other organs of plants also contain yellow pigments. These are seldom apparent except in leaves in which the chlorophyll fails to develop, or in which it is destroyed as a result of senescence or other physiological changes. Corn plants that have grown from seed in the dark, for example, do not synthesize chlorophyll, but are usually yellow in colour because of the presence of yellow pigments.

In the autumn disintegration of the chlorophyll in the leaves of many woody species results in unmasking the yellow pigments which are also present in the leaves. The yellow chloroplast pigments of leaves belong to the group of compounds called *carotenoids.* Except in blue-green algae and the photosynthetic bacteria, chlorophylls occur only in the chloroplasts. The carotenoids of leaves are also restricted to the chloroplasts, and this also appears to be true of the carotenoids in the cells of most algae. All of these pigments appear to be present only in the grana of the chloroplasts. In certain other plant organs, such as flower petals,

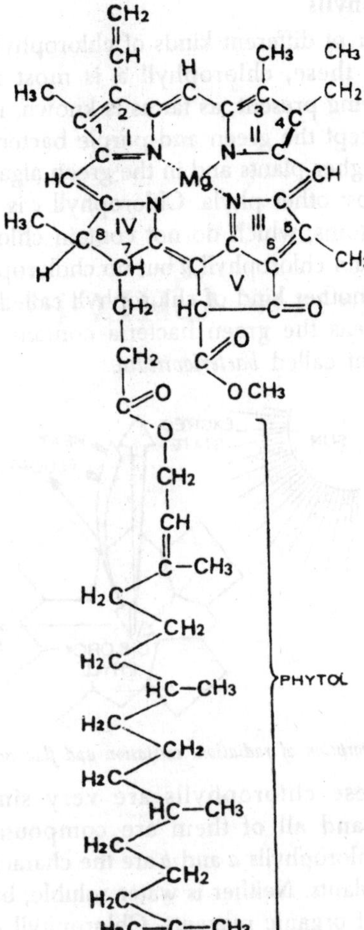

Fig. 4.1. Structure of chlorophyll a.

the yellow pigments occur in chromoplasts. The chlorophylls and associated carotenoids are often called the *chloroplast pigments.* In some kinds of algae pigments of another group, called the *phycobilins,* also occur in the chloroplasts. The best known of these are the *phycoerythrin* of most red and some blue-green algae, and the *phycocyanin* of most blue-green and some red algae. These two pigments are closely related chemically. Both are colloidal proteinaceous pigments and can be extracted from the cells with hot water.

The Chlorophylls

A number of different kinds of chlorophyll occur in the plant kingdom. Of these, chlorophyll a is most nearly of universal occurrence being present, as far as is known, in all photosynthetic organisms except the green and purple bacteria. Chlorophyll *b* is found in all higher plants and in the green algae, but is not present in algae of most other phyla. Chlorophyll *c* is found in the brown algae and diatoms, which do not contain chlorophyll *b*. Similarly red algae contain chlorophyll *d* but no cholorophyll *b*. In the purple bacteria still another kind of chlorophyll called *bacteriochlorophyll* is present, whereas the green bacteria contain another apparently similar pigment called *bacterioviridin.*

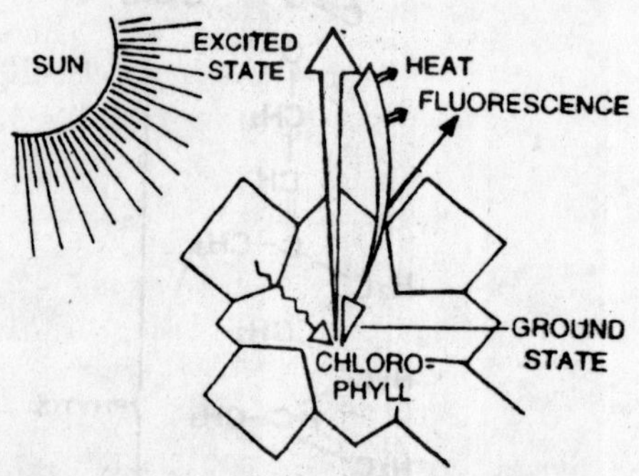

Fig. 4.2. Absorption of radiation, excitation and fluorescence by chlorophyll.

All of these chlorophylls are very similar in chemical composition, and all of them are compounds which contain magnesium. Chlorophylls *a* and *b* are the characteristic chlorophylls of the higher plants. Neither is water soluble, but both are soluble in a number of organic reagents. Chlorophyll *a* is readily soluble in absolute-ethyl alcohol, ethyl ether, acetone, chloroform, and carbon bisulfide. Chlorophyll *b* is soluble in the same reagents, although generally less so.

Chloroophyll *a* is blue-green in solution, and blue-black in the solid state; chlorophyll *b* is almost pure green in solution, and greenish-black in the solid state. All of the chlorophylls possess the property of *fluorescence.* This term refers to the peculiar property possessed by certain substances when illuminated of re-radiating light of wave lengths other than those absorbed by the substance. Usually the radiated light (fluorescent light) is longer in wave length than the incident light. Chlorophyll *a* in ethyl alcoholic solution

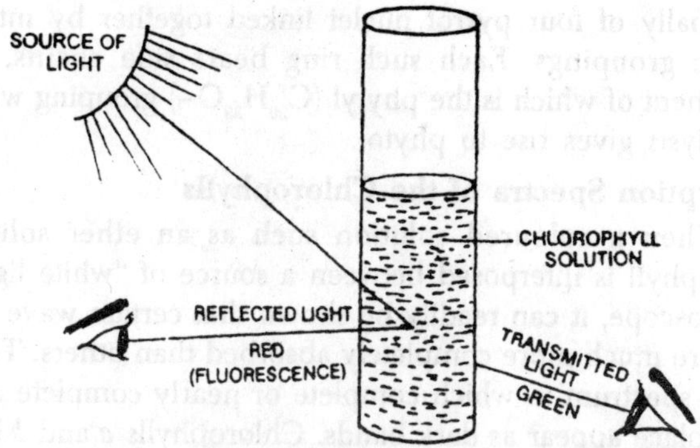

Fig. 4.3. Appearance of chlorophyll in transmitted and reflected light.

exhibits a deep blood red fluorescence, best seen by viewing the solution in reflected light. Similar solutions of chlorophyll *b* exhibit a brownish-red fluorescence. Chlorophyll in living cells also exhibits fluorescence.

The Chemistry of the Chlorophylls

Willstatter and his associates have isolated chlorophylls *a* and *b* in pure form from over two hundred different species of higher plants and found them to be identical in chemical composition. They also succeeded in determining the molecular formulas of these two chlorophylls. For chlorophyll *a* this is $C_{55}H_{72}O_5N_4Mg$; for chlorophyll *b* it is $C_{55}H_{70}O_6N_4Mg$. Each chlorophyll yields a separate series of degradation products, but one of the products derived from both after mild hydrolysis is an unsaturated primary alcohol which is called *phytol* ($C_{20}H_{39}OH$).

Splitting of the phytol group from the chlorophyll molecule can also be accomplished by the enzyme *chlorophyllase*, which occurs in leaves. Phytol makes up about one-third of the chlorophyll molecule. It has a strong affinity for oxygen and may be responsible for the reducing action of chlorophyll. When ashed pure chlorophyll leaves a residue composed solely of magnesium oxide. Although iron and other minerals seem essential for the formation of chlorophyll in living cells magnesium is the only metallic constituent of the chlorophyll molecule. From the results of the intensive study of the chemistry of the degradation products of chlorophyll the probable structural formulas of both chlorophyll *a* and chlorophyll *b* have been determined. The nucleus of a molecule of chlorophyll *a* consists of a complex ring structure composed principally of four pyrrol nuclei linked together by intermediate atomic groupings. Each such ring bears side chains, the most prominent of which is the phytyl ($C_{20}H_{39}O-$) grouping which upon hydrolysis gives rise to phytol.

Absorption Spectra of the Chlorophylls

When a coloured solution such as an ether solution of a chlorophyll is interposed between a source of "white light" and a spectroscope, it can readily be shown that certain wave lengths of light are much more completely absorbed than others. The regions of the spectrum in which complete or nearly complete absorption takes place appear as dark bands. Chlorophylls *a* and *b* both show a definite and characteristic absorption spectrum when in solution. The exact appearance of such absorption spectra will vary, however, depending upon the kind of solvent and the concentration and thickness of the layer of solution which is examined. Absorption curves for chlorophyll *a* and *b* in either solution as determined by photoelectric measurements. Both chlorophylls exhibit maximum absorption in the blue-violet region and a secondary maximum in the short red. As a rule the absorption spectrum of a leaf does not correspond very closely with the spectrum of a solution of the pigments from that same leaf.

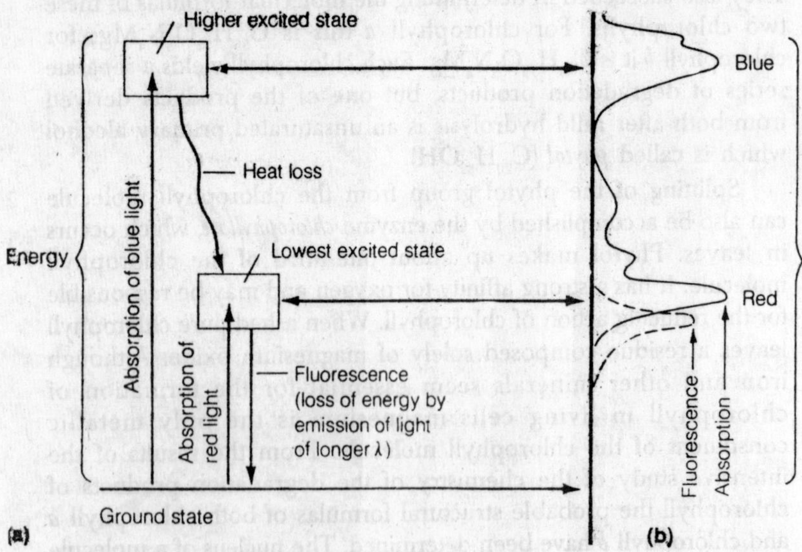

Fig. 4.5. Light absorption and emission by chlorophyll. Part (a) shows an energy level diagram. (b) The spectra of absorption and fluorescence.

In general there is relatively more absorption of light by leaves in those spectral regions where absorption by solutions is low, and a relatively less absorption of light by leaves in those spectral regions where absorption by solutions is high. Some of the principal reasons for this are : (1) the pigment in the leaf is not continuous, but located only in the plastids; hence some light may pass through the leaf without traversing any pigment; (2) actual or apparent absorption, the latter a result of scattering of light within the leaf, by the non-pigmented parts of cells affects the overall absorption of light by the leaf; and (3) the concentration of the pigment in the leaf is usually different from its concentration in extracts; as a result specific absorption values of pigments in the leaf are different from those in a solution.

Physicochemical State of the Chlorophylls in the Chloroplasts

Although the extensive investigations of extracts of chlorophylls in organic solvents have advanced our knowledge of their physical and chemical properties, it should be remembered that these are only *extracts* and the physicochemical state of the chloropohylls in them bears little or no relation to its condition in the chloroplasts. One necessary step in learning how the chlorophylls of living cells operate will be to discover how they are related physically and chemically to the other constituents of the chloroplasts. Any acceptable picture of the physicochemical relationships of the chlorophylls within the cholorpolasts must also be reconcilable with the known structural arrangement of the chlorophylls in the grana of the chloroplasts.

A number of investigators have shown that green, opalescent, aqueous extracts can be prepared from the leaves of many species by grinding them with sand in water or solutions and clarifying by filtration, precipitation, or centrifugation. These extracts possess many of the properties of protein sols or suspensions and the chlorophyll molecules seem to be definitely associated with the protein micelles or particles. These results have led to the postulation of the existence of a chlorophyll-protein compound of constant composition in plants, analogous to the hemoglobin (hemin plus globin, a protein) of the blood. Although results of these experiments indicate that some sort of an association of chlorophylls with proteins in the chloroplasts is highly probable, there are strong reasons for doubting that this association takes the form of a protein-chlorophyll compound in the strictly chemical sense.

Actual chemical analyses of the chloroplast substance have been made by Chibnall (1939), Menke (1940), Granick (1938), Comar (1942), and others. The Protein content of the chloroplast material from several species has been found to lie within a range of 35 to 55 per cent, and the lipid content within a range of 18 to 32 percent on the dry weight basis (chloroplasts are probably about 50 percent water). In the leaf cells of spinach most of the lipids present are constituents of the chloroplasts. In such chemical analyses the chlorophylls and carotenoids (see later) constitute a part of the lipid fraction, about 5-10 per cent of which is chlorophylls and 2-4 per cent carotenoids. The relatively large quantity of lipids in the chloroplasts suggests that the pigments commonly present in some manner may be associated with lipids as well as with proteins. The physicochemical properties of the lipids are such that it seems more probable that any association between them and the chlorophylls or carotenoids is of a physical rather than a chemical nautre.

Chlorophyll Synthesis

Chlorophylls, in common with practically all the other organic substances which occur in plants, are products of the synthetic activities of the plant. A number of conditions are known to be necessary for or at least to influence greatly, the synthesis of chlorophylls in plants. Absence of any one of these factors will inhibit chlorophyll synthesis resulting in the condition often called *chlorosis.* This term is most frequently applied when the failure of chlorophylls to develop is the result of a deficiency of one of the essential mineral elements. Different types of chlorosis may develop in the leaves of any species depending upon the factor limiting chlorophyll formation. The following discussion of the factors influencing chlorophyll synthesis refers primarily to the formation of chlorophylls *a* and *b* in the higher plants.

Genetic factors

That certain genetic factors are necessary for the development of chloropohylls is shown by the behaviour of some varieties of maize in which a certain proportion of the seedlings produced cannot synthesize chloropohylls, even if all environmental conditions are favourable for their formation. As soon as the food stored in the grain is exhausted such "albino" seedlings die. This trait is inherited in such strains of maize as a Mendelian recessive and hence is apparent only in plants homozygous for this factor. More

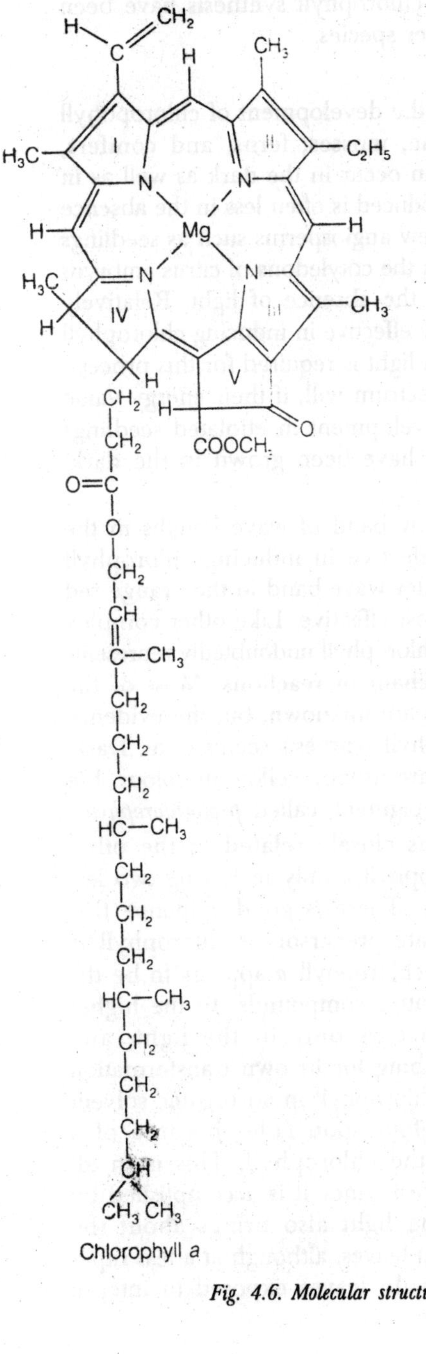

Chlorophyll *a*

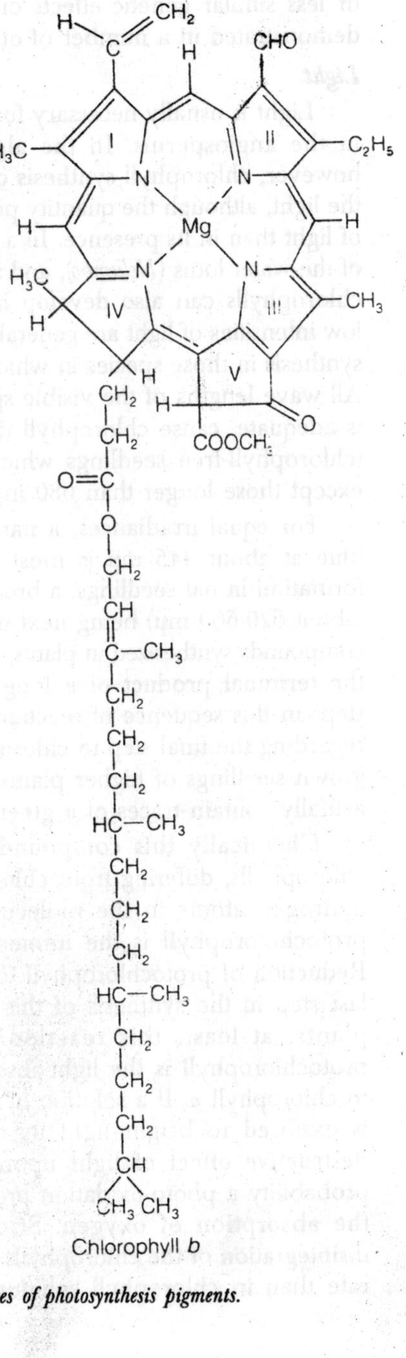

Chlorophyll *b*

Fig. 4.6. Molecular structures of photosynthesis pigments.

or less similar genetic effects on chlorophyll synthesis have been demonstrated in a number of other species.

Light

Light is usually necessary for the development of chloropohyll in the angiosperms. In the algae, mosses, ferns, and conifers, however, chlorophyll synthesis can occur in the dark as well as in the light, although the quantity produced is often less in the absence of light than in its presence. In a few angiosperms such as seedlings of the water lotus (*Nelumbo*), and in the cotyledons of citrus embryos chlorophylls can also develop in the absence of light. Relatively low intensities of light are generally effective in inducing chlorophyll synthesis in those species in which light is required for this process. All wave lengths of the visible spectrum will, if their energy value is adequate, cause chlorophyll development in etiolated seedlings (chlorophyll-free seedlings which have been grown in the dark) except those longer than 680 mμ.

For equal irradiances, a narrow band of wave lengths in the blue at about 445 mμ is most effective in inducing chlorophyll formation in oat seedlings, a broader wave band in the orange red (about 620-660 mμ) being next most effective. Like other complex compounds synthesized in plants, chlorophyll undoubtedly represents the terminal product of a long chain of reactions. Most of the steps in this sequence of reactions are unknown, but the evidence regarding the final step in chlorophyll synthesis seems clear. Dark-grown seedlings of higher plants are usually yellow in colour, but actually contain traces of a green pigment, called *protochlorophyll.*

Chemically this compound is closely related to the other chlorophylls, differing from chlorophyll *a* only in having two less hydrogen atoms in the molecule. There is good evidence that protocholorophyll is the immediate precursor of chlorophyll *a.* Reduction of protochlorophyll to chlorophyll *a* appears to be the last step in the synthesis of this latter compounds. In the higher plants, at least, this reaction occurs only in the light, and protochlorophyll is the light-absorbing for its own transformation to chlorophyll *a.* If a solution of chlorophyll in an organic solvent is exposed to bright light its colour soon fades because of a destructive effect of light upon the chlorophyll. This is in all probability a photo-oxidation process, since it is accomplished by the absorption of oxygen. Strong light also brings about the disintegration of the chlorophylls in leaves, although at a less rapid rate than in chlorophyll solutions. In leaves exposed to intense light, therefore, synthesis and decomposition of chlorophylls are probably going on simultaneously. In accord with this concept are the results of Shirley (1929) who found in a number of species of plants that the chlorophyll content per unit leaf weight or per unit leaf area increased with decreasing light intensity until a relatively low intensity was reached. Further decrease in light intensity below this value caused a decrease in chlorophyll content.

Oxygen

In the absence of oxygen etrolated seedlings fail to develop chlorophylls even when illuminated under conditions otherwise favourable for chlorophyll formation. Oxygen must therefore be required in some of the steps in the sequence of reactions whereby chlorophyll is synthesized.

Carbohydrates

Etiolated leaves which have been depleted of soluble carbohydrates fail to turn green even when all of the other conditions to which they are exposed favour chlorophyll synthesis. When such leaves are floated on a sugar solution, sugar is absorbed and chlorophyll formation occurs rapidly. A supply of carbohydrate foods is therefore essential for the formation of chlorophyll.

Nitrogen

Since nitrogen is a part of chlorophyll molecules it is not surprising to find that a deficiency of this element in the plant retards chlorophyll formation. Failure of chlorophylls to develop is one of the commonly recognized symptoms of nitrogen deficiency in plants.

Magnesium

Like nitrogen this element is also a part of chlorophyll molecules. Deficiency of magnesium in plants results in the development of a characteristic mottled chlorosis of the older leaves.

Iron

In the absence of iron in an available form green plants are unable to synthesize chlorophylls and the leaves soon become balanced or yellow in colour. While not a constituent of chlorophyll molecules, iron is essential for its synthesis.

Other Mineral Elements

In the absence of manganese, copper, or zinc, more or less characteristic chloroses develop in plants. These substances apparently play at least indirect roles in the synthesis of chlorophyll.

Temperature

In general, chlorophyll synthesis can occur over a wide range of temperatures. In etiolated wheat plants synthesis of chlorophylls occurs at any temperature within the range 3°-48°C, but is most rapid between 26° and 30°C. Somewhat similar results have been obtained on chlorophyll synthesis in potato tubers except that synthesis appears to be most rapid in the approximate range 11°-19°C. The range of temperatures over which chlorophyll synthesis occurs as well as the temperatures of most rapid synthesis undoubtedly varies considerably from one species to another.

Water

Desiccation of leaf tissues not only inhibits synthesis of chlorophylls but seems to accelerate disintegration of the chlorophylls already present. A familiar example of this effect is the browning of grass during droughts.

The mechanism of chlorophyll synthesis is very sensitive to any type of physiological disturbance within the plant. Many other conditions besides those already discussed, such as lack of certain other mineral elements, deficient aeration of the roots, infestations of insects, or infection with bacterial, fungous or viral diseases, may induce,directly or indirectly, partial or complete chlorosis of the leaves. The failure of chlorophyll synthesis to take place normally is often one of the first observable symptoms of almost any derangement in the metabolic conditions within a plant.

The Carotenoids

This group of orange, red, yellow and brownish pigments includes a number of chloroplast pigments. At least sixty different carotenoids are known to be present in plants. Best known of the pigments in this group is the orange-yellow *carotene.* This is a hydrocarbon with the formula $C_{40}H_{56}$ and exists in three isomeric forms β-carotene is the most abundant of these in plants, and appears to be invariably present in the chloroplasts. This compound is the precursor of vitamin A which is formed in the animal body by a simple hydrolytic splitting of the molecules of β-carotene, as follows:

$$C_{40}H_{56} + 2H_2O \rightarrow 2C_{20}H_{29}OH$$

Lycopene, a red pigment found in the fruits of tomato, red peppers, roses, and other species, is also isomeric with carotene.

The *xanthophylls* or *carotenols* are mostly yellow or brownish pigments, many, but not all, of which have molecular formula

$C_{40}H_{56}O_2$. Chemically the carotenols are closely related to carotene,and undoubtedly transformations from one kind of carotenoid to another occur readily in plants.

Luteol is by far the most abundant carotenol is leaves other leaf xanthophylls include zeaxanthol, violaxanthol, cryptoxanthol, flavoxanthol, and neoxanthin.

Numerous other xanthophylls occur in the various phyla of algae. Among the most abundant of these pigments are the fucoxanthins which impart to brown algae their distinctive colour. Several brownish carotenols are also present in the diatoms, including fucoxanthins, diatoxanthin, and diadinoxanthin.

The carotenoids are not water soluble, but they can be extracted from plant tissues by use of suitable organic solvents. Carotene, for example, is soluble in ehtyl ether, chloroform, and carbon bisulfide. Leaf xanthophylls are soluble in chloroform and ethyl alcohol, but only slightly soluble in carbon bisulfide.

All of the carotenoids are highly unsaturated compounds, and hence readily oxidizable. Loss of a considerable proportion of the carotenoids present in a plant tissue may be occasioned by oxidation during extraction procedures.

As shown by its absorption spectrum β-carotene absorbs only wave lengths in the blue-violet portion of the visible spectrum. The absorption spectra of other carotenoids are similar to that of carotene.

Most, if not all, of the carotenoid pigments can be synthesized by plants in the absence of light.

Relative Quantities of the Chloroplast Pigments Present in Green Leaves

Data regarding the molecular proportions of the chloropolast pigments in the leaves of several species are presented in Table 3.1. The leaves listed in this table contain about three molecules of the chlorophylls to one molecule of the carotenoids, about three molecules of chlorophyll *a* to each molecule of chlorophyll *b*, and about two molecules of xanthophylls to each molecule of carotene. Although these values can be taken as representative, considerable variation in the proportions of the chloroplast pigments occurs as a result of differences in genetic factors, environmental conditions, and the age of the leaf. As shown in Table 4.1, with senescence of leaves in the autumn there is a marked increase in the proportion of carotenoids to chlorophylls as a result of the distintegration of the chlorophylls. The proportion of chlorophyll *a* to *b* in leaves may vary considerably depending upon the light intensity to which the leaves have been exposed. Willstatter and Stoll (1928) and others, for example, have shown that the proportion of chlorophyll *b* and *a* is higher in shade leaves of many species than in sun leaves of the same species. As an example of genetic differences in the pigment content of leaves, results obtained with dark and burley varieties of tobacco may be taken as illustrative. Although the ratio of chlorophyll *a* to *b* was virtually the same in all varieties studied the total chlorophyll content per gram of dry weight was, on the average,more than 50 percent greater in the dark varieties than in the burley varieties. Carotene content of the dark varieties was also appreciably higher than in the burley varieties.

Table 4.1. Molecular proportions of the chloroplast pigments present in leaves at different seasons.

Species	*Date*	*Appearance of leaves*	*Molecular proportions* — *Chlorophyll a* / *Chlorophyll b*	*Carotene* / *Xanthophylls*	*Chlorophyll a and b* / *Carotene and Xanthophylls*
European Elder	July	Green	2.74	0.59	3.33
(*Sambucus nigra*)	Oct. 21	Green	2.61	0.60	3.15
Sunflower	Oct. 7	Green	2.86	0.54	3.66
Helianthus annuus	Nov. 5	Yellow green	4.01	0.26	1.27
	Nov. 5	Yellow	0.95	0.17	0.32
Horse Chestnut	Oct. 24	Green	4.23	0.47	3.06
(*Aesculus hippo-castanum*)	Oct. 24	Yellow green	4.23	0.20	0.81

THE ROLE OF CHLOROPHYLL IN PHOTOSYNTHESIS

Any effort to understand the basis of life on this planet must always come back to photosynthesis: the process that enables plants to grow by utilizing carbon dioxide (CO_2), water (H_2O) and a tiny amount of minerals. Photosynthesis is the one large-scale process that converts simple, stable, inorganic compounds into the energy rich combination of organic matter and oxygen and thereby makes abundant life on earth possible. Photosynthesis is the source of all living matter on earth, and of all biological energy.

The overall reaction of photosynthesis can be summarized in the following equation : $CO_2 + H_2O + \text{light} \rightarrow (CH_2O) + O_2 +$ 112,000 calories of energy permole. (CH_2O) stands for a carbohydrate; for example, glucose: $(CH_2O)_6$. "Mole" is short for "gram molecule" : one gram multiplied by the molecular weight of the substances in question–in this case carbohydrate and oxygen.

The state of knowledge of photosynthesis was first summarized 30 years ago. The whole process was heavily shrouded in fog (see "Photosynthesis," by Eugene I. Rabinowitch; Scientific American Offprint 34). Later investigation had penetrated the mists sufficiently to disclose some of the main features of the process (see "Progress in Photosynthesis," by Eugene I. Rabinowitch; Scientific American, November, 1953). Since then much new knowledge has been accumulated in particular the sequence of chemical steps that convert carbon dioxide into carbohydrate is now understood in considerable detail. The fog has thinned out in other areas, and the day when the entire sequence of physical and chemical events in photosynthesis will be well understood seems much closer.

The photosynthetic process apparently consists of three main stages:(1) the removal of hydrogen atoms from water and the production of oxygen molecules; (2) the transfer of the hydrogen atoms from an intermediate compound in the first stage to one in the third stage, and (3) the use of the hydrogen atoms to convert carbon dioxide into a carbohydrate.

The least understood of these three stages is the first : the removal of hydrogen atoms from water with the release of oxygen. All that is known is that is entails a series of steps probably requiring several enzymes, one of which contains manganese. The third stage–the production of carbohydrates from carbon dioxide–is the best understood, thanks largely to the work of Melvia Calvin and his co-workers at the University of California at Berkeley. The subject of our article is the second stage : the transfer of hydrogen atoms from the first stage to the third. This is the energy-storing part of photosynthesis; in it, to use the words of Robert Mayer, a discoverer of the law of the conservation of energy, "the fleeting sun rays are fixed and skillfully stored for future use."

The light energy, to be converted into chemical energy by photosynthesis is first taken up by plant pigments, primarily the green pigment chlorophyll. In photosynthesis chlorophyll functions as a photocatalyst; when it is in its energized state, which results from the absorption of light it catalyzes an energy-storing chemical reaction. This reaction is the primary photochemical process; it is followed by a sequence of secondary "dark"–that is, nonphotochemical–reactions in which no further energy is stored. Once it was thought that in photosynthesis the primary photochemical process is the decomposition of carbon dioxide into carbon and oxygen, followed by the combination of carbon and water.

More recently it has been suggested that the energy of light serves primarily to dissociate water, presumably into hydroxyl radicals (OH) and hydrogen atoms; the hydroxyl radicals would then react to form oxygen molecules. It is better than either of these two formulations to say that the primary photochemical process in photosynthesis is the boosting of hydrogen atoms from a stable assocation with oxygen in water molecules to a much less stable one with carbon in organic matter. The oxygen atoms "left behind" combine into oxygen molecules, an association also much less stable than the one between oxygen and hydrogen in water. The replacement of stable bonds (between oxygen and hydrogen) by looser bonds (between oxygen and oxygen and between hydrogen and carbon) obviously requires a supply of energy, and its explains why energy is stored in photosynthesis.

The transfer of hydrogen atoms from one molecule to another is called oxidation-reduction. The hydrogen atom is transferred from a donor molecule (a "reductant") to an acceptor molecule (an "oxidant"); after the reaction the donor is said to be oxidized and the acceptor to be reduced. The transfer of an electron can often substitute for the transfer of a hydrogen atom: in an aqueous system (such as the interior of the living cell) there are always hydrogen ions (H^+), and if such anion combines with the electron becomes equivalent to the acquisition of a hydrogen atom (electron + H^+ ion → H^+ atom).

The chain of oxidation-reduction reactions in photosynthesis has some links that involve electron transfers and others that involve electron transfers and others that invlove hydrogen-atom transfers. For the sake of simplicity we shalls speak of electron transfers, with the understanding that in some cases what is actually transferred is a hydrogen atom. Indeed, the end result of the reactions undoubtedly is the transfer of hydrogen atoms.

In the oxidation-reduction reactions of photosynthesis the electrons must be pumped "uphill", that is why energy must be supplied to make the reaction go. The tiny chlorophyll containing chloroplasts of the photosynthesizing plant cell act as chemical pumps; they obtain the necessary power from the absorption of light by chlorophyll (and to some extent from absorption by other pigments in the chloroplast). It is important to realize that the energy is stored in the two products organic matter and free oxygen and not in either of them separately. To release the energy by the combustion of the organic matter (or by respiration, which is slow, enzyme-catalyzed combustion) the two products must be brought together again. How much energy is stored in the transfer of electrons from water to carbon dioxide, converting the carbon dioxide to carbohydrate and forming a proportionate amount of oxygen ?

Oxidation-reduction energy can conveniently be measured in terms of electrochemical potential. Between a given donor of electrons and a given acceptor there is a certain difference of oxidation- reduction potentials. This difference depends not only on the nature of the two reacting substances but also on the nature of the products of the reaction ; it is characteristic of the two oxidation-reduction "couples". For example, when oxygen is reduced to water (H_2O) its potential is +.81 volt, but when it is reduced to hydrogen peroxide (H_2O_2) the potential is +.27 volt.. The more positive the potential, the stronger is the oxidative power of the couple; the more negative the potential, the stronger is its reducing power.

When two oxidation-reduction couples are brought together, the one containing the stronger oxidant tends to oxidize the one containing the stronger reductant. In photosynthesis, however, a weak oxidant (CO_2) must oxidize a weak reductant (H_2O), producing a strong oxidant (O_2) and a strong reductant (a carbohydrate). This calls for a massive investment of energy. The specific amount needed is given by the difference between the oxidation-reduction potentials of the two couples involved in the reaction: oxygen-water and carbon-dioxide- carbohydrate. The oxygen -water potential, is about +.8 volt; the carbon-dioxide-carbohydrate potential, about –.4 volts. The transfer of a single electron from water to carbon dioxide thus requires +.8 minus –.4 or 1.2, electron volts of energy. For a molecule of carbon dioxide to be reduced to CH_2O–the elementary molecular group of a carbohydrate–for electrons (or hyrogen atoms) must be transferred; hence the total energy needed is 4.8 electron volts. This works out to 112,000 calories per mole of carbon dioxide reduced and of oxygen liberated.

In short, the pumping of electrons in the second stage of photosynthesis entails the storage of 112,000 calories of energy per mole for each set of four electrons transferred. We know the identity of the primary electron donor in photosynthesis (water) and of the ultimate electron acceptor (carbon dioxide), but what are the intermediates involved in the transfer of electrons from the first stage to the third? This has become the focal problem in recent studies of the photosynthesis process.

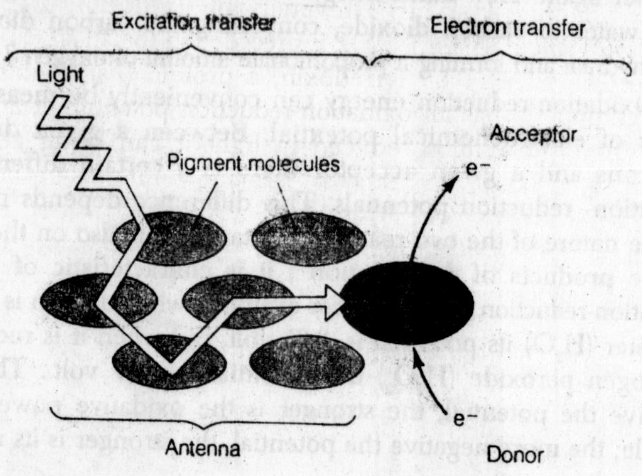

Fig. 4.7. The basic concept of energy transfer during photosynthesis.

As a matter of fact, it is not yet definitely known what compound releases electrons from the first stage and what compound receives them in the third; that is why these compounds are respectively labeled ZH and X in the illustration at the left. About the donor, ZH, we have almost no information; the following considerations suggest the possible nature of the primary acceptor, X. From the study of the mechanism of respiration we are familiar with an important oxidation-reduction catalyst: nicotinamide adenine dinucleotide phosphate, or NADP (formerly known as triphosphopyridine nucleotide, or TPN). NADP has an oxidation reduction potential of about–.32 volt, thus in itself it is not a strong enough reductant to provide the–.4 electron volt potential needed to reduce carbon dioxide to carbohydrate. NADP can achieve this feat, however, if it is supplied with additional energy in the form of the high-energy compound adenosine triphosphate, or ATP.

A molecule of ATP supplies about 10,000 calories per mole when its terminal phosphate group is split off, and this is enough to provide the needed boost to the reducing power of NADP. Furthermore, we know that NADP is reduced when cell-free preparations of chlcroplasts are illuminated. Put together, these two facts led to the now widely accepted hypothesis that the second stage of photosynthesis manufactures both ATP and reduced NADP and feeds them into the third stage. At first it was assumed that NADP is identical with X, the primary acceptor in our scheme. Subsequent experiments by various workers–notably Anthony San Pietro at Johns Hopkins University of California at Berkley–suggested, however, that NADP is preceded in the "bucket brigadge" of electron transfer by ferredoxin, a protein that contains iron. This compound has an oxidation-reduction potential of about–.42 volt; therefore if it is reduced in light it can bring about the reduction of NADP by a "dark" reaction requiring no additional energy supply.

More recently Bessel Kok of the Research Institute for Advanced Studies in Baltimore has found evidence suggesting that compound X may be a still stronger reductant, with a potential of about– .6 volt. If this is so, plants have the alternatives of either applying this stronger reductant directly to the reduction of carbon dioxide or letting it reduce first ferredoxin and then NADP and using reduced NADP to reduce carbon dioxide. It seems a round about procedure to create a reductant sufficiently strong for the task at hand, then to sacrifice a part of its reducing power and finally to use ATP to compensate for the loss. It is not unknown, however, for nature to resort to devious ways in order to achieve its aims.

For photosynthesis to be a self-contained process the required high energy phosphate ATP must be itself manufactured by photosynthesis. The formation of ATP has in fact been detected in illuminated fragments of bacteria by Albert W. Frenkel of the University of Minnesota and in chloroplast fragments by Arnon and his co-workers. As a matter of fact, November, 1960. As a matter of fact, ATP is needed not only to act as a booster in the reduction of an intermediate in the carbon cycle by reduced NADP but also for another step in the third stage of photosynthesis. According to a sequence of reactions worked out in 1951 by Andrew A. Benson and his colleagues at the University of California at Berkeley, carbon dioxide enters photosynthesis by first reacting with a "carbon dioxide acceptor", a special sugar phosphate called rebulose diphosphate. It turns out that the production of this compound from its precursor-ribulose monophosphate–cells for a molecule of ATP.

ATP is produced both in chloroplasts and in mitochondria, the tiny intracellular bodies that are the site of the energy-liberating stage of respiration in animals as well as plants. The mitochondria produce ATP as their main function, exporting it as packaged energy for many life processes. The chloroplasts, on the other hand, make ATP only as an auxiliary source of energy for certain internal purposes. The energy of the light falling on the chloroplasts is stored mostly as oxidation- reduction energy by the uphill transfer of electrons. Only a relatively small fraction is diverted to the formation of ATP, and this fraction too ultimately becomes part of the oxidation-reduction energy of the final products of photosynthesis: oxygen and carbohydrate.

Let us now consider the uphill transport of electrons in greater detail. Recent investigations have yielded considerable information about this stage. Apparently the pumping of the electrons is a two-stop affair, and among the most important intermediates in it are the catalysts called cytochromes.

The idea of a two-step electron transfer process grew from a consideration of the energy economy of photosynthesis. Precise measurements, particularly those made by the late Robert Emerson and his co-workers at the University of Illinoise, showed that the reduction of one molecule of carbon dioxide to carbohydrate, and the liberation of one molecule of oxygen, requires a minimum of eight quanta of light energy. The maximum quantum yield of photosynthesis, defined as the number of oxygen molecules that can be released for each quantum of light absorbed by the plant cell, is thus 1/8, or 12 percent. Since the transfer of four electrons is involved in the reduction of one carbon dioxide molecule, it was suggested that it takes two light quanta to move each electron.

Emerson and his colleagues went on to determine the quantum yield of photosynthesis in maximicronatic right of different wavelengths throughout the visible spectrum. They found that the yield, although is remained constant at about 12 percent in most of the spectrum, dropped sharply near the spectrum's far-red end. This decline in quantum yield, called the "red drop," begins at a wavelength of 680 millimicrons in green plants and at 650 millimicrons in red algae.

These are two chlorophylls present in green plants ; chlorophyll *a* and chlorophyll *b*. Only chlorophyll *a* absorbs light at wavelengths longer than 680 millimicrons; the absorption of chlorophyll *b* rises to a peak at 650 millimicrons and becomes negligible at about 680 millimicrons. Emerson found that the quantum yield of photosynthesis at the farred end of the spectrum beyond 680 millimicrons can be brought to the full efficiency of 12 percent by simultaneously exposing the plant to a second beam of light with a wavelength of 650 millimicrons. In other words, when light primarily absorbed by chlorophyll *a* was supplemented by light primarily absorbed by chlorophyll *b*, both beams gave rise to oxygen at the full rate. This relative excess in photosynthesis when a plant is exposed to two beams of light simultaneously, as compared with the yield produced by the same two beams separately, is known as the Emerson effect, or enhancement.

On the basis of his discovery Emerson concluded that photosynthesis involves two photochemical processes : one using energy supplied by chlorophyll *a*, the other using energy supplied by chlorophyll *b* or some other "accessory" pigment. Experimenting with various combinations, of a constant far-red beam with beams of shorter wavelength, and using four different types of algae (green, red, blue-green and brown), Emerson's group found that the strongest enhancement always occurred when the second beam was absorbed mainly by the most important accessory pigment (the green pigment chlorophyll *b* in green cells, the red pigment phycoerythrin in red algae, the blue pigment phycocyanin in blue-green algae and the reddish pigment fucoxanthol in brown algae). Such results suggested that these other pigments are not mere accessories of chlorophyll *a* but have an important funciton of their own in photosynthesis.

Certain findings concerning the behaviour of pigments in living plant cells, however, seemed to make this conclusion untenable, illuminated plant cells fluoresce; that is pigment molecules energized by the absorption of light quanta reemit some of the absorbed energy as fluorescent light. The source of fluorescence can be identified, because each substance has its own characteristic fluorescence spectrum. The main fluorescing pigment in plants always proves to be chlorophyll *a*, even when the light is absorbed by another pigment. This had first been shown for brown algae in a study conducted in 1943 by H.J. Dutton, W.H. Manning and B.B. Duggar at the University of Wisconsin; later the finding was extended to other organisms by L.N.M. Duysens of the University of Leiden. Known as sensitized fluorescence, the phenomenon indicates that the initial absorber has transferred its energy of excitation to chlorophyll *a*; the transfer is effected by a kind of resonance process. Careful measurements have shown that certain accessory pigments–chlorophyll *b*, phycoerythrin, phycocyanin, and fucoxanthol–pass on to chlorophyll *a* between 80 and 100 percent of the light quanta they absorb. For some other accessory pigments–for example carotene–the transfer is less efficient.

This puts accessory pigments back in the role of being mere adjuncts to chlorophyll *a*. True, they can contribute, by means of resonance transfer, light energy to photosynthesis, thereby improving the supply of energy in regions of the spectrum where chlorophyll *a* is a poor absorber. Chlorophyll *a*, however, collects all this energy before it is used in the primary photochemical process. Why, then, the enhancement effect ? Why should chlorophyll *a* need, in order to give rise to fullrate photosynthesis, one "secondhand" quantum obtained by resonance transfer from another absorber in addition to the one quantum it had absorbed itself?

A better understanding of this paradox resulted from the discovery that there apparently exist in the cell not only chlorophyll *a* and chlorophyll *b* but also two forms of chlorophyll *a*. These two forms have different light-absorption characteristics, and they probably also have different photochemical functions.

In the living cell chlorophyll *a* absorbs light most strongly in a broad band with its peak between 670 and 680 millimicrons. In our laboratory at the University of Illinois we undertook to plot the Emerson effect more carefully than before as a function of the wave length of the enhancing light. We found that for green and brown cells the resulting curve showed, in addition to peaks corresponding to strong absorption by the accessory pigments, a peak at 670 millimicrons that must be due to chlorophyll *a* itself. It was this finding that suggested the existence of two forms of chlorophyll *a*. The form that absorbs light at the longer wavelengths–mainly above 680 millimicrons–seemed to belong to one pigment system, now often called System I. The form that absorbs at 670 millimicrons seemed to belong to another pigment system; System II. In the second system the form of chlorophyll *a* that absorbs at 670 millimicrons is strongly assisted by accessory pigments, probably by resonance transfer of their excitation energy. Careful analysis of the absorption band of chlorophyll *a* by C.Stacy French of the Carnegie Institution of Washington's Department of Plant Biology, and also in our laboratory, confirmed that the band is double, with one peak near 670 millimicrons (at 668 millimicrons) and another band at 683 millimicrons.

If chlorophyll *a* is extracted from living plants, there is only one product; we must therefore assume that in the living cell the two forms differ in the way molecules of chlorophyll *a* are clumped together, or in the way they are associated with different chemical partners (proteins, lipids or other substances). Be this as it may, the important implication of the new finding is that photosynthesizing cells possess two light-absorbing systems,one containing a form of chlorophyll *a* absorbing around 683 millimicrons and the other a form absorbing around 670. The latter system includes chlorophyll *b* (in green-plant cells) or other accessory pigment (in brown, red and blue-green algae).

Further investigation–particualrly of red algae–has suggested however that the distribution of these two components in the two systems may be less clear-cut. In red algae a large fraction of the chlorophyll *a* absorbing at 670 millimicrons seems to belongs to System I rather than System II. In all likelihood the two systems provide energy for two different photochemical reactions, and efficiency in photosynthesis requires that the rates of the two reactions be equal. What are these reactions? This question brings us to another significant finding, which suggested the participation of cytochromes in photosynthesis.

Cytochromes are proteins that carry an iron atom in an attached chemical group. They are found in all mitochondria, where they serve to catalyze the reactions of respiration. Robert Hill and his co-workers at the University of Cambridge first found that chloroplasts also contain cytochromes–two kinds of them. One, which they named cytochrome *f*, has a positive oxidation-reduction potential of about 4 volt. The other, which they name cytochrome b_6, has a potential of about 0 volt. In 1960 Hill, together with Fay Bendall, proposed an ingenious hypothesis as to how the two cytochromes might act as intermediate carriers of electrons and connect the two photochemical systems. They suggested that cytochrome b_6 receives an electron by a photochemical reaction from the electron donor ZH; the electron is then passed on to cytochrome *f* by a "downhill' reaction requiring no light energy. (The oxidation-reduction potential of cytochrome *f* is much more positive than that of cytochrome b_6). A second photochemical reaction moves the electron uphill again, from cytochrome *f* to the electron-acceptor X in the third stage of photosynthesis. In this sequence the photochemical reactions store energy and the reaction between the two cytochromes releases energy.

Some of the released energy, however, can be salvaged by the formation of an ATP molecule: this occurs in the transfer of electrons among cytochromes in respiration. In this way ATP is obtained without spending extra light quanta on its formation, which the tight energy economy of photosynthesis does not allow. Experiments by Duysens and his associates confirmed this hypothesis, by showing that the absorption of light by System I causes the oxidation of a chytochrome, whereas the absorption of light by System II causes its reduction. This is exactly what we would expect. The illustration on the opposite page shows that the light reaction of System II should flood the intermediates between the two photochemical reactions with electrons taken from ZH; the light reaction of System I should drain these electrons away, sending them up to the acceptor X and into the third stage of photosynthesis. This, then, describes in a general way the oxidation-reduction process by which the chloroplasts store the energy of light in photosynthesis.

Several other investigators have contributed evidence for the two-step mechanism; notable among them are French, Kok, Arnon, Horst Witt of the Max Vollmer Institute in Berlin and their colleagues. In detail the process probably is much more complex than our scheme suggests. Its "downhill" central part seems to include, in addition to the two cytochromes,certain compounds of the group known as quinones and also plastocyanin, a protein that contains copper. What is known of the submicroscopic structure in which the reactions of the second stage of photosynthesis take place?

There is much evidence that the photosynthetic apparatus consists of "units" within the chloroplast, such unit containing about 300 chlorophyll molecules. This picture first emerged from experiments conducted in 1932 by Emerson and William Arnold on photosynthesis during flashes of light; it was later supported by various other observations. The pigment molecules are packed so closely in the unit that when one of them is excited by light it readily transfers its excitation to a neighbour by resonance. The energy goes on travelling through the unit, rather as the steel ball in a pinball machine bounces around among the pins and turns on one light after another. Eventually the migrating energy quantum arrives at the entrance to an enzymatic "conveyor belt", where it is trapped and utilized either to load an electron onto the belt or to unload one from it. (The steel-ball analogy should not be taken litorally the migration of energy is a quantum-mechanical phenomenon, and the quantum's location can only be defined in terms of probability; its entrapment depends on the probability of finding it at the entrance to the conveyor belt). How is the quantum trapped?

The trap must be a pigment molecule with what is called a lower excited state; the migrating quantum can stumble into such a molecule but cannot come out of it. Kok has found evidence that System I contains a small amount of a special form of chlorophyll called pigment 700 because it absorbs light at a wavelength of 700 millimicrons; this pigment could serve as a trap for the quantum bouncing around in System I. There seems to be a proper amount of pigment 700 : about one molecule per unit. Furthermore, experiments suggest that pigment 700 is oxidized by light absorbed in System I and reduced by light absorbed in System II. It has an oxidation reduction potential of about +.4 volt.

All these properties fit the role we have assigned pigment 700 in our scheme : collecting energy from a 300 molecule unit in System I, using it to transfer an electron to the acceptor X and recovering the electron from cytochrome *f*. One suspects that there should be a counterpart of pigment 700 in System II, but so far none has been convincingly demonstrated. We believe, however, that a pigment we have tentatively named pigment 680–from the anticipated position of its absorption band–does serve as an energy trap in System II. Its existence is supported by the discovery of a new fluorescent emission band of chlorophyll at 693 millimicrons, which is compatible with absorption at 680 millimicrons. This band is emitted by certain algae when they are exposed to strong light of the wavelengths absorbed by System II. What is the spatial organization of the pigment systems in the electron-boosting mechanism of the second stage of photosynthesis?

It seems that the two systems may be arranged in two monomolecular layers, with a protein layer between them containing the enzymatic conveyor belt. Chloroplasts are known from electron microscope studies to consist of a set of lamellae: thin alternating layers of protein and fatty material piled one atop the other. Each layer appears to consist of particles arrayed rather like cobblestones in a pavement. The particles were first observed in electron micrographs made by E. Steinmann of the Technische Hochschule in Zurich; subsequently Roderio B. Park and John Biggins of the University of California at Berkeley made clearer micrographs of the particles and named them quantasomes.

The units comprising Systems I and II may operate independently or they may be sufficiently close together to exchange energy by resonance, when such exchange is needed to maintain a balanced rate of operation by the two systems. The picture of the energy-storing second stage of photosynthesis presented in this article is, of course, still only a working hypothesis. Alternative hypotheses are possible, one of which we shall briefly describe. For many years the late James Franck, who shared the Nobel prize in physics for 1925, tried to develop a plausible physio-chemical mechanism of photosynthesis. In 1963 he proposed, together with Jerome L. Rosenberg of the University of Pittsburgh, a concept according to which the two consecutive photochemical stops occur in one and the same energy trap.

In other words, according to Franck, the same chlorophyll molecule that takes the electron away from the initial donor ZH and transfer of the electron from the cytochrome to the acceptor X. In the first transfer, Franck suggested, the chlorophyll a molecule functions in the short-lived "singlet "excited state (in which the valence electrons have opposite spins); in the second transfer it functions in the long-lived "triplet" state (in which the valence electrons have parallel spins). Franck's hypothesis avoids certain difficulties of the "two trap" theory, but new difficulties arise in their place. On balance the two-trap picture seems to us the more plausible one at present. No doubt this picture will change as more information emerges. It is merely a first effect to penetrate the inner sanctum of photosynthesis, the photocatalytic laboratory in which the energy of sunlight is converted into the chemical energy of life.

5

Factors Affecting Photosynthesis

Earlier investigators of the effects of various conditions upon the rate of photosynthesis attempted to distinguish among *minimum*, *optimum*, and *maximum* values for each factor in relation to photosynthesis. In evaluating the effect of temperature upon photosynthesis, for example, it was generally considered that there was a minimum temperature below which on photosynthesis occurred, an optimum at which the process takes place most rapidly, and a maximum above which photosynthesis ceases. Advocates of this point of view, however, soon found themselves confronted with the anomalous situation of a fluctuating "optimum."

The "optimum" carbon dioxide concentration was found to be greater at high light intensities than at low ones, the "optimum" temperature was found to vary with the light intensity, and the "optimum" light intensity was different for plants well supplied with water than for those which were inadequately supplied. The first important step in the classification of this problem of the influence of various factors upon photosynthesis was taken when Blackman (1905) enunciated the "principle of limiting factors." This principle is essentially an elaboration of Liebig's "law of the minimum" and was stated by its author as follows: "When a process is conditioned as to its rapidity by a number of separate factors, the rate of the process is limited by the pace of the 'slowest' factor". The explanation of this principle can best be presented in terms of the illustration given by Blackman. Assume the intensity of light

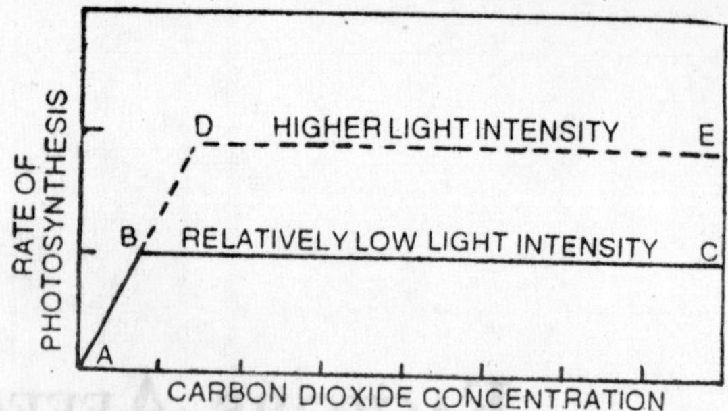

Fig. 5.1. Diagram to illustrate Blackman's interpretation of the principle of limited factors.

to be just great enough to permit a leaf to utilize 5 mg. of carbon dioxide per hour in photosynthesis. If only 1 mg. of carbon dioxide can enter the leaf in an hour, the rate of photosynthesis is limited by the carbon dioxide factor.

As the carbon dioxide supply is increased the rate of photosynthesis is also increased until 5 mg. of carbon dioxide enter the leaf per hour. Any further increase in the supply of carbon dioxide will have no influence upon the rate of photosynthesis, unless a sufficient concentration is present to bring about retarding effects, because insufficient light energy is available to permit its utilization. Light has now become the limiting factor and further increase in the rate of photosynthesis can be brought about only by an increase in the intensity of light. This theory assumes a progressive increase in the rate of the process with a quantitative increase in the limiting factor (in this example, carbon dioxide) until the point is reached at which some other factor becomes limiting (in this example, light intensity).

At this point the increase stops abruptly, and the rate of photosynthesis becomes constant. According to this concept, when the magnitude of photosynthesis is limited by one of a set of factors, only a shift in that factor toward a condition more favourable for the process will result in an increase in the rate of photosynthesis. If the light intensity is sufficient to permit the leaf to utilize 10 mg. of carbon dioxide per hour, then the rate of photosynthesis will rise with increase in the carbon dioxide concentration to a value about twice as great as that at which the maximum rate of photosynthesis was attained at the lower light intensity.

Light and carbon dioxide are not the only factors which can be limiting in the photosynthetic process. Theoretically, as examples given later in the chapter will show, any of the factors which influence this process can, under certain conditions, become, limiting.

Most subsequent workers have been unable to accept the principle of limiting factors in quite the simple form in which it was first proposed by Blackman. Most investigators have found that, when the rate of increase of photosynthesis is plotted along the ordinate with the quantitative variations in some one factor as the abscissa, the resulting curve is not found to show an abrupt transition to the horizontal as postulated by Blackman's formulation of this principle, but shows instead a gradual transition to a position approximately parallel to the abscissa. Within this transition region it is evident that in rease in either of the two factors involved will result in an increase in the rate of photosynthesis.

The explanation for the occurrence of this gradual transition in the direction of the curve, rather than the abrupt change postulated by the original Blackman theory, results at least in part from the fact that the seat of the photosynthesis is in the chloroplasts of which there are millions in even a small leaf. Patently it is impossible that each and every chloroplast will be subjected to exactly the same conditions at exactly the same time. All of the chloroplasts are not equally exposed to light; neither are all of them equally well supplied with carbon dioxide. As a factor approaches a limiting value it may check the rate of photosynthesis in some chloroplasts sooner than in others. It is quite possible therefore for light to be the limiting factor for some chloroplasts, while carbon dioxide is simultaneously the limiting factor for other chloroplasts. Similar comments apply to most of the other factors influencing photosynthesis. Hence the rate of photosynthesis, as measured in terms of entire organs, will exhibit only a gradual change when factors affecting the process are modified, and there exist well-defined regions in curves, in which two or even more factors may be considered to act simultaneously as limiting factors.

It should be clearly understood that in speaking of "limiting factors" as applied to physiological processes, it is not their absolute magnitude which is significant, but their relative magnitude in proportion to the amounts actually required in the process. The quantitatively smallest factor does not necessarily condition the rate of the process, since it may be necessary only in traces, while larger amounts of some other material, or a greater intensity of some other factor may be necessary. To illustrate, suppose that we assume ten units of *a*, two units of *b*, and one unit of *c* are necessary for the formation of one unit of *d*. If we suppose that only five units of *a* are available, none of *d* can be formed regardless of the quantities of *b* and *c* available. Although *c* may be in absolute minimum, *a* is in relative minimum and thus acts as the limiting factor. For this reason many authorities consider it to be more satisfactory to speak of the "relatively limiting factor," "factor in relative minimum," or "most significant factor," rather than of the "limiting factor".

One other qualification of the theory of limiting factors not envisaged in the original Blackman formulation should also be mentioned. When present in a high enough intensity or concentration, most of the factors influencing the rate of photosynthesis exert a depressing or inhibitory effect on the process.

The modifications which have been imposed upon the original concept of limiting factors do not invalidate this principle as a good approximation to the facts, nor do they destroy its value as a point of view from which to interpret the influence of various factors upon the rate of photosynthesis. The significant fact is that the rate of the process, except in relatively narrow transition regions, is usually determined in the main by the least favourable factor, which may for convenience be spoken of either at the limiting factor or as the factor in relative minimum.

The Role of Carbon Dioxide

All of the carbon dioxide used by green plants reaches the chloroplasts as dissolved carbon dioxide, carbonic acid, or one of the salts of the latter. In land plants the atmosphere is the only important source of carbon dioxide. Carbon dioxide released in respiration may be utilized in photosynthesis without leaving the plant, but under conditions favourable for photosynthesis this does not constitute a very large fraction of the total used. Submersed water plants use carbonates and bicarbonates as well as dissolved carbon dioxide and carbonic acid as substrates of photosynthesis. The atmosphere is composed chiefly of two gases, nitrogen (about 78 percent) and oxygen (about 21 percent), but also contains, in addition to a variable but never large amount of water-vapour, small quantities of other gases. One of its minor constituents, carbon dioxide, which constitutes on the average only about 0.03 percent by volume of the atmosphere, plays a role of the greatest

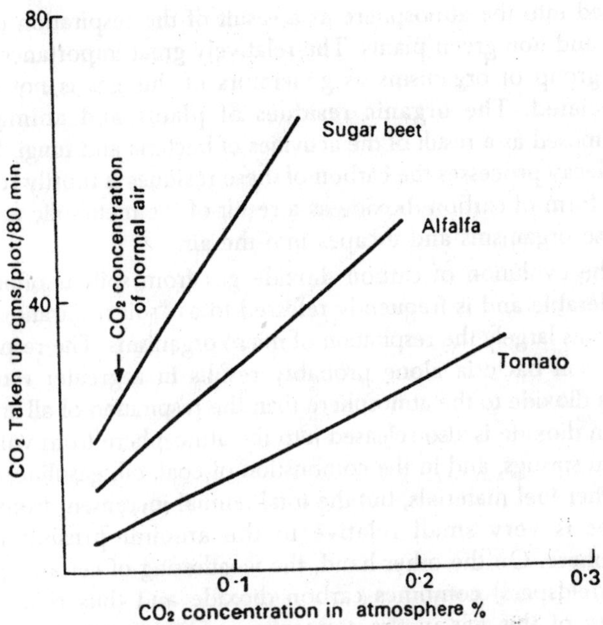

Fig. 5.2. Effect of CO_2 concentration on the rate of photosynthesis.

significance in the biological world. As a result of the photosynthesis activity of green plants, the carbon dioxide from the air becomes chemically bound for periods of indefinite length in the organic molecules which are the basis of all life.

In view of its important biological role, the proportion of carbon dioxide in the atmosphere seems precariously small. The absolute amount present, however, is enormous. Estimates, necessarily very approximate, place the total quantity of carbon dioxide in the atmosphere at about 2×10^{15} kg. According to Schroeder (1919), the quantity of carbon dioxide used annually in photosynthesis by land plants is about 6×10^{13} kg. This is only about one-thirtieth the amount actually present in the atmosphere.

Sources of the Atmospheric Carbon Dioxide

While green plants are continually removing carbon dioxide from the atmosphere, other processes are continually replenishing the atmospheric reservoir with this gas. Carbon dioxide is continually being returned to the atmosphere as a product of the respiration of plants and animals. Plants are more important producers of carbon dioxide than animals. Carbon dioxide is released into the atmosphere as a result of the respiration of both green and non-green plants. The relatively great importance of the latter group of organisms as generators of this gas is not always appreciated. The organic residues of plants and animals are decomposed as a result of the activities of bacteria and fungi. During such decay processes the carbon of these residues is mostly released in the form of carbon dioxide as a result of the metabolic activities of these organisms and escapes into the air.

The evolution of carbon dioxide gas from soils is often very considerable and is frequently referred to as "soil respiration"; this represents largely the respiration of micro-organisms. The respiration of the soil bacteria alone probably results in a greater return of carbon dioxide to the atmosphere than the respiration of all animals. Carbon dioxide is also released into the atmosphere from volcanos, mineral springs, and in the combustion of coal, oil, gasoline, wood, and other fuel materials, but the total annual increment from these sources is very small relative to the amount present in the atmosphere. On the other hand, the weathering of certain igneous locks (feldspars) combines carbon dioxide and thus reduces the quantity of this gas in the atmosphere. Oceans are much more important reservoirs of carbon dioxide than the atmosphere.

The oceans occupy nearly three fourths of the earth's surface and are estimated to contain about eighth times as much carbon in forms available to plants as the atmosphere. The carbon dioxide in ocean waters is involved in a complex series of chemical and biological cycles which have never been fully evaluated. Marine plants consume carbon dioxide in photosynthesis and release it in respiration. Marine animals feed either upon marine plants or other animals, but ultimately, as with land animals, all of their food comes from the process of photosynthesis. A part of the carbon in the food consumed by such organisms is released into the water as carbon dioxide in the process of respiration.

Aquatic micro-organisms accomplish the decay of dead plants and animals, releasing most of the carbon in these organic remains in the form of carbon dioxide. Complex equilibria between the dissolved carbon dioxide, carbonates and bicarbonates also exist. Large quantities of carbonates are tied up by certain marine animals in the formation of shells. Other marine animals precipitate large quantities of carbon dioxide in chemically combined form as the calcium carbonate of calcareous rocks. The conversion of

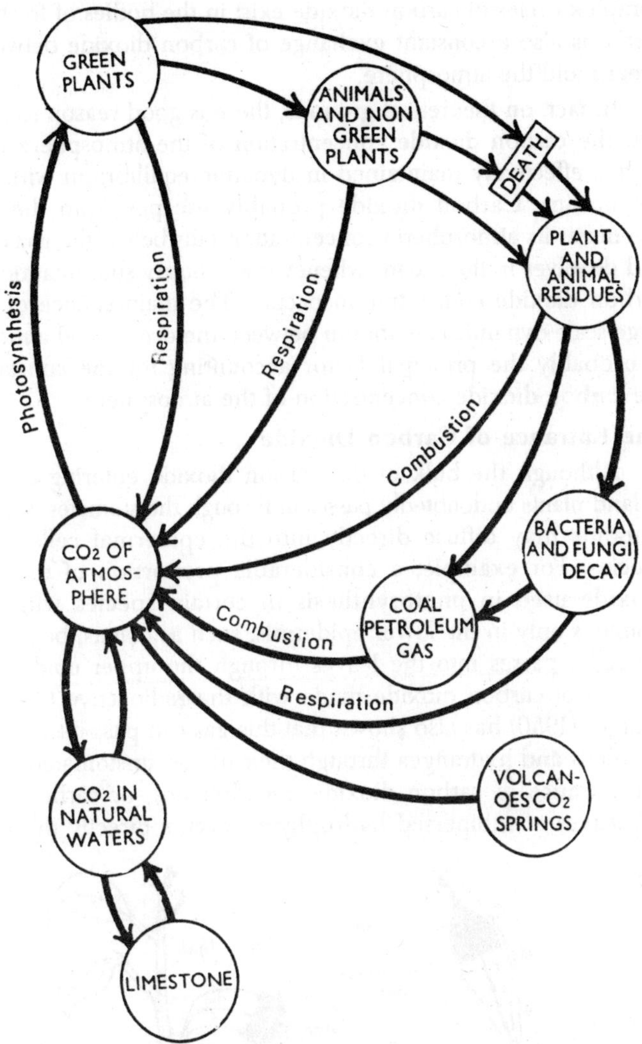

Fig. 5.3. Carobon dioxide cycle in nature.

bicarbonates into carbonates results in the release of carbonic acid and thus increases the available carbon dioxide content of the water. Eventually such rocks (limestones, etc.) may be raised above sea level, and the carbon dioxide tied up in the form of carbonates again is released to the atmosphere or dissolved in running water during dissolution of the rock. Similar although not quite such complex cycles of carbon dioxide exist in the bodies of fresh water. There is also a constant exchange of carbon dioxide between the oceans and the atmosphere.

In fact, on theoretical grounds, there is good reason to suppose that the carbon dioxide concentration of the atmosphere is more or less effectively maintained in dynamic equilibrium with that of the oceans. Carbon dioxide probably escapes from the oceans whenever its atmospheric concentration falls below the usual value, and dissolves in the oceans whenever a contrary shift in atmospheric carbon dioxide concentration occurs. The maintenance of such a large-scale dynamic equilibrium between the oceans and atmosphere is probably the principal factor accounting for the constancy of the carbon dioxide concentration of the atmosphere.

The Entrance of Carbon Dioxide

Although the bulk of the carbon dioxide entering the leaves of land plants undoubtedly passes in through the stomates, substantial amounts may diffuse directly into the epidermal cells in some species. For example, a considerable proportion of the carbon dioxide used in photosynthesis in certain species which have stomates only in the lower epidermis, such as coleus, begonia, and avocado, passes into the leaves through the upper epidermis. By the use of carbon dioxide made with the radioactive C^{14} isotope. Dugger (1950) has also shown that this gas can pass into the leaves of coleus and hydrangea through their upper, unstomated surfaces. All entrance of carbon dioxide, bicarbonates, or carbonates into the leaves of submersed hydrophytes, occurs directly through the

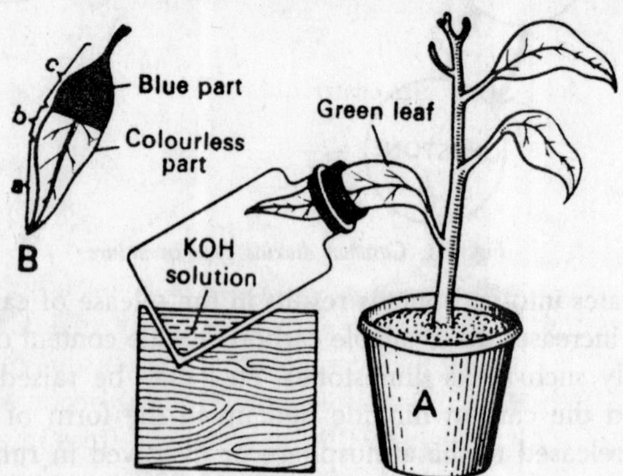

Fig. 5.4. Moll's half leaf experiment to demonstrate that CO_2 is essential for photosynthesis.

epidermis. The rate of entrance of carbon dioxide through stomates is very considerable in proportion to the aggregate area of the stomatal pores.

Under conditions favourable for photosynthesis Brown and Escombe (1900) found that carbondioxide would diffuse into a catalpa leaf from the atmosphere at the rate of 0.07cc. per cm^2 of leaf surface per hour. Since the stomates in this leaf occupy only 0.9 percent of the surface, diffusion of carbon dioxide gas through them took place at a rate of 7.77cc. per cm^2 of stomatal aperture per hour. (The assumption was made that all carbon dioxide entered through the stomates). Under the same conditions a normal secution of sodium hydroxide absorbs from the atmosphere, even in rapidly moving air, only 0.177cc. of carbon dioxide per cm^2 per hour. In other words, carbon dioxide gas diffuse through the stomates at a rate approximately fifty times as fast as it diffuses into an efficient absorbing surface.

The extremely rapid rate of diffusion of carbon dioxide gas through the stomates can be interpreted in the light of principles of diffusion of gases through small openings. The work of Brown and Escombe, indicates that these same principles hold, in general, for carbon dioxide as well as water-vapour.

The most important principle illustrated, that the rate of diffusion does not decrease proportionately with reduction in the aggregate area of pores, is shown in the last two columns. For septum 4 (pores 10.52 diameters apart), for example, the diffusion is 31.4 percent of that from, the open tube, although only 0.70 percent of the septum surface is occupied by the apertures.

Effects of Variations in the Atmospheric Concentration of Carbon Dioxide upon the Rate of Photosynthesis

Although the average carbon dioxide content of the atmosphere is 0.03 percent, the actual range in plant habitats is from about half to several times this value. Lower than average values are not infrequent in relatively quiet air in zones in which high rates of photosynthesis prevail. In a field of corn, for example, the carbon dioxide content of the air surrounding the plants may become measurable less than the average atmospheric value during daylight hours whenever high rates of photosynthesis occur. Similar, in the atmosphere on the level with the crowns of trees in a dense forest the carbon dioxide content in sometimes considerably less than the average atmospheric value during hours when rapid photosynthesis is in progress. In a tightly closed greenhouse the carbon dioxide content of the air may decrease materially during the course of a bright day.

Investigations of Lundegardh (1931) and others indicate that a considerable part of the carbon dioxide utilized by plants in many habitats is released locally as a result of "soil respiration," i.e., in the respiration of soil microganisms. Such a release of carbon dioxide is especially pronounced in well fertilized soils, soils rich in organic matter, and in many forest soils. "Soil respiration," when marked, may result in a local enrichment of the carbon dioxide concentration in the air stratum close to the surface of the ground. Such a rise in carbon dioxide content is greatest during the night hours when the off setting effect of photosynthesis in low-growing plants is absent. On a well-fertilized field the formation of carbon dioxide as a result of "soil respiration" during a 24-hr, period may equal or exceed the consumption in photosynthesis during the daylight hours.

The carbon dioxide content of the air is also influenced by fogs and mists. The quantity present per unit volume of air is measurably higher on foggy mornings than on clear mornings, and rates of apparent photosynthesis are correspondingly increase in a foggy as contrasted with a clear atmosphere if no other condition are limiting.

Although it is customary to express atmospheric concentrations of carbon dioxide as percentage by volume, such units are valid only when all of the concentrations to be compared are at approximately the same atmospheric pressure. The rate of diffusion of carbon dioxide into a leaf is a function of its partial pressure which varies directly as the total atmospheric pressure. Hence, although the percentage of carbon dioxide in the atmosphere is about 0.03 percent both at sea level and, for example, at an altitude of 15,000 feet, its partial pressure is very different. At sea level the partial pressure of this gas is about 0.23 mm. Hg; at 15,000 feet altitude, about 0.13 mm, Hg. In terms of fundamental units, therefore, the effective concentration of carbon dioxide varies considerably with altitude, and this may be a factor to be taken into consideration when photosynthetic rates or development of plants at different altitudes is being compared.

In general, at least over relatively short periods, with increase in the concentration of atmospheric carbon dioxide, there is an increase in the rate of photosynthesis until some other factor, most commonly light, becomes limiting. The results of an experiment on the relation of carbondioxide concentration to the rate of photosynthesis, in which, it should be noted, the highest light intensity used was only about 10 percent of noonday summer sunlight.

If no other factor is limiting , the rate of photosynthesis rises in most species, at least for a while, with increase in the carbon dioxide concentration up to, at a minimum, 15-20 times the usual atmospheric concentration. Relatively high concentrations of carbon dioxide, in general, retard photosynthesis. The concentration at which such depressant effects are initiated varies with the kind of plant, stage of development of the photosynthesis tissue, length of exposure to carbon dioxide, and other prevailing environmental conditions. In the leave of *Hydrangea otaksa*, for example, the rate of photosynthesis, although considerably retarded, is not entirely inhibited in short time experiments at concentrations as great as 20 percent carbon dioxide.

Under natural conditions during summer months in moist, temperate regions the carbon dioxide concentration of the atmosphere is most frequently the limiting factor in photosynthesis for all photosynthetic tissues which are well exposed to light. There seems to be little doubt that increase in the atmospheric concentration of this gas to at least several times its average value of 0.03 per cent has a continuously favourable effect on photosynthesis, as long as no other factors are limiting. There is evidence, however, that only slightly higher concentrations of carbon dioxide which favour enhanced photosynthesis over periods of a few hours or days may have a detrimental effect on the process over longer periods. Thomas and Hill (1949), for example, found that tomato plants exposed to ten times the atmospheric concentration of carbon dioxide during daylight hours showed detrimental effects within less than 2 weeks.

The Role of Light

The energy stored by green plants in the molecules of carbohydrates during photosynthesis can be supplied only by light. Any source of radiant energy which includes wave lengths within the range of the visible spectrum will induce photosynthesis, provided its intensity is sufficiently great. Although a few of the longer wavelengths of ultraviolet apparently are effective in photosynthesis, and some of the shorter wave lengths of infrared can be used by sulphur bacteria, in general this process can occur only in the visible part of the spectrum. Under natural conditions sunlight, either direct or reflected from the sky or other objects, is the only source of radiant energy including wave lengths which can be used in photosynthesis.

Photosynthesis will occur under electric lights or other artificial sources of illumination if of sufficient intensity. Such light sources are often used in experimental work on photosynthesis and to some extent in greenhouses as supplementary sources of illumination. Light, like all forms of radiant energy, varies in intensity (irradiance), quality and duration and the influence of this factor upon photosynthesis will be discussed under these three headings. Before the various effects of light upon photosynthesis are considered, however, it will be desirable to analyze the physics relations between leaves and incident light.

The Optical Properties of Leaves

Of the visible light which falls on leaves, as with radiant energy generally, a part is reflected, a part is transmitted through the leaf, and a part is absorbed by the leaf. The proportion of the *visible*

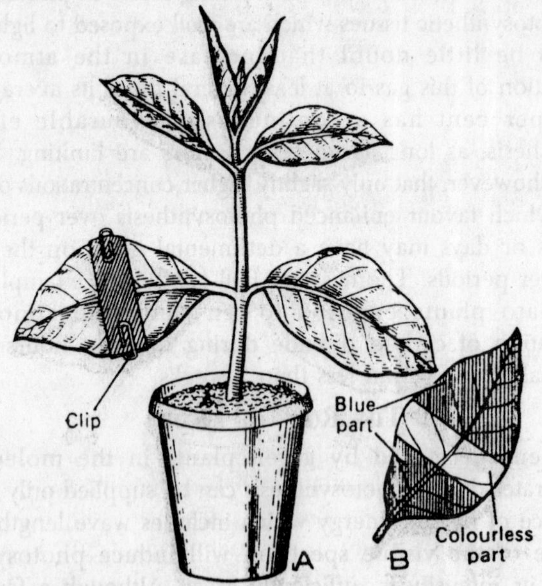

Fig. 5.5. Experiment to show that light is essential for photosynthesis.

light incident upon a leaf which is a absorbed varies considerably according to the kind of leaf and the intensity of light, but it is frequently in the neighbourhood of 80 percent and probably often higher. A clear distinction should be drawn between the proportion of visible light absorbed by a leaf and the proportion of the total incident radiant energy absorbed, the latter being in the neighbourhood of 50 percent for many kinds of leaves. For any given kind of leaf the proportions of incident light reflected, transmitted, and absorbed vary considerably according to wave length.

In general a larger proportion of the incident light is absorbed by thick than by thin leaves. For normally green leaves maximum reflection in the range of visible light occurs in the extreme long red with a secondary, but less pronounced, maximum in the green. Conversely, maximum absorption is in the blue-violet and orange-short red regions, corresponding approximately with the prominent absorption bands of the chlorophylls. Reflection of infrared by leaves is about 50-60 per cent in the wave length range of 760-1100 mμ.

Effects of Variations in the Intensity of Light upon the Rate of Photosynthesis

In general, with increase in light intensity there is an increase in the rate of photosynthesis until some other factor, in this example, the carbon dioxide concentration, becomes limiting. At relatively low light intensities, as long as carbon dioxide is not the limiting factor, the rate of photosynthesis as approximately proportional to the light intensity. The maximum light intensity used was not great enough for the carbon dioxide factor to have become limiting. It should also be noted that the maximum light intensity employed–1000 foot candles–is much inferior in intensity to usual summer sunlight, which at noon on a clear day usually is equivalent to 8,000-10,000 foot-candles. At usual atmospheric carbon dioxide concentrations, maximum photosynthetic rates are attained when leaves are exposed to light intensities considerably less than the maximum sun light intensity.

In apple leaves for example, peak photosynthetic a rates are reached at light intensities one-fourth to one-third full sunlight. Similarly in maize leaves maximum rates are attained at about one-fourth the peak intensity of summer sunlight and in loblolly pine (*Pinus taeda*) needles about one-third full sunlight. On the other hand, in pronounced shade species, the highest rates of

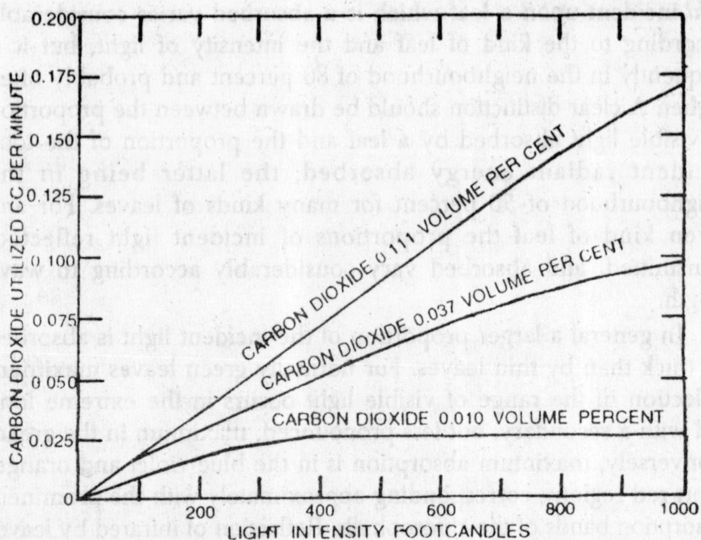

Fig. 5.6. Relation between different light intensities and rate of photosynthesis of wheat plants at three different carbon dioxide concentrations.

photosynthesis under otherwise favourable natural conditions are reached at light intensities no higher than 1000 foot-candles. Results such as those discussed above will be obtained only when a single leaf or a small plant, in which there is little or no shading of one part by another, is used as the experimental material. When the effect of light on photosynthesis is considered in terms of an entire tree, however, a different relation holds.

Thus Heirecke and Childers (1937) showed that the rate of photosynthesis for an entire apple tree increased progressively with increase in light intensity up to or nearly that of full sunlight. This is undoubtedly because many of the interior leaves on a large tree are heavily shaded. Although, in general, maximum rates of photosynthesis are attained in the leaves of most (probably all) species at light intensities considerably below that of full sunlight, these investigators showed that many of the interior leaves of an apple tree receive as little as 1 percent or less of the sunlight received by peripheral leaves. Even in full sunlight, therefore, many of the leaves on an apple tree do not photosynthesize at their maximum capacity. The lower the light intensity the greater the proportion of the leaves of which this will be true. Hence the greater the intensity of the incident light, the greater the average rate of photosynthesis per unit of lead area. The total photosynthesis per tree, therefore, increases progressively with increased illumination up to or at least close to the maximum possible sunlight intensity.

Internal shading effects on the rate of photosynthesis have been demonstrated experimentally for such diverse plants as the loblolly pine (*Pinustaeda*) by Kramer and Clark (1947) and the hornwort (*Ceratophyllum demersum*)–a submersed aquatic–by Meyer (1939) and doubtless occur in many other kinds of plants with densely arranged foliage. Similar effects exist in compact stands of plants such as fields of some crop species. Mutual shading of the plants in plots of alfalfa is sufficiently marked so that the photosynthetic a rate of such plots considered as a unit is highest at peak light intensities, or at intensities close thereto, as in an apple tree.

In fields of maize, on the other hand, the plants apparently do not shade each other sufficiently to have any appreciable effect on the aggregate rate of photosynthesis on bright days. The intensity of the sunlight incident on the earth's surface varies from hour to hour and from season to season, as well as with meterological conditions. Clouds, fogs, dust, and atmospheric humidity, all influence the intensity of the radiation which reaches the surface of the earth. The exposure and pitch of a slope are also factors influencing the intensity of the light impinging upon a given location, and are particularly of importance in hilly or mountainous country. Other conditions being equal the intensity of sunlight also increases with increase in altitude. In aquatic habit at slight intensity decreases with depth below the surface of the water. Most variations in the intensity of natural light are accompanied by variations in light quality of greater or lesser magnitude. Usually, however, under out-of-doors conditions, variations in the intensity component of light are of greater physiological influence than the accompanying variations in the quality component.

Ecologically one of the most important factors causing different species of plants to be exposed to differences in light intensity is the effect of taller plants in shading those of lesser stature. Some species thrive and photosynthesize efficiently only in fully exposed locations; others can complete their normal life cycle in intensely shaded habitats. Even under a tree with a rather open crown the light intensity is only one-tenth to one-twentieth that of full sunlight. Hence light is usually the limiting factor for photosynthesis in most species of plants when growing in the shade of trees. Species which normally grow in deep shade are often exceptions to this statement.

On cloudy days light is also usually the limiting factor in photosynthesis. Because of the prevalence of cloudy weather in many regions during the winter, plants under glass often photosynthesize at very low rates during those months. The short length of the daylight period also contributes to low daily rates of photosynthesis during the winter season. A number of studies have been made of the minimum light intensities at which various species of plants are just barely able to survive. Unless foods have previously accumulated, the minimum light intensity at which a plant will remain alive for indefinite periods during its normally active seasons must permit sufficient photosynthesis to compensate for both day and night respiration, and in all probability must also allow for some assimilation.

Bates and Roeser (1928) studied the effects of low light intensities upon the growth of a number of species of evergreens native to the western United States. The seedlings were exposed to a 200 watt, blue tungsten lamp for 10 hours a day. Differences in light intensity were obtained by growing the seedlings at different distances from the source of illumination. Their results show that redwood seedlings were able to maintain their initial dry weight in a light intensity less than 1 percent of full sunlight, while pinon pine required about 6 percent, and the other three species were intermediate in their requirement. Extremely high light intensities exert an inhibitory effect on photosynthesis, a phenomenon called *solarization.*

Holman (1930) showed that bean leaves exposed to an illumination of 6800 foot-candles accumulated starch readily, but that little starch was synthesized at twice this illumination value. This effect is exerted upon the photosynthetic mechanism itself and not on the process of starch synthesis. Solarization effects appear to result principally, and probably entirely, from the phenomenon of *photo-oxidation*, in which leaves consume oxygen in the light, and use it is the oxidation of certain cell constituents, carbon dioxide being released in the process.

The occurrence of photo-oxidation in *Chlorella*, but the existence of this phenomenon has also been demonstrated in other species. At lower light intensities a steady or nearly study rate of photosynthesis is maintained by *Chlorella* but the higher the light intensity above 4000 foot-candles the more markedly the rate falls off with time. At 27,700 foot candles, the highest intensity uses, oxygen consumption (photo-oxidation) sets in almost immediately. Photo-oxidation should not be confused with ordinary respiration as it is an entirely different process. It operates through at least a part of the photosynthetic mechanism and may occur at a rate three or four times greater than respiration.

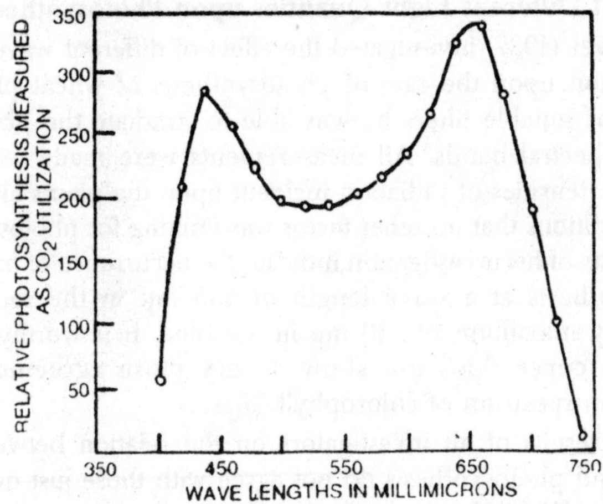

Fig. 5.7. Relative rate of photosynthesis of wheat in different wavelengths of light of equal intensity.

In a sense photo-oxidation may be regarded as a variant of photosynthesis in which oxygen is used as the substrate instead of carbon dioxide. Oxygen consumption at highlight intensities results not merely from a super imposing of photo oxidation upon photosynthesis but at least partly from an inactivation of the latter process by the former. Short periods of photo-oxidation are usually not harmful to leaves or other photosynthetic organs, but continuation for this process for more than a few hours commonly results in the decolorization of chlorophylls and ultimately in death of the cells in which it is occurring. Light may exert indirect as well as direct effects upon photosynthesis.

Low light intensities favour stomatal closure and hence may sometimes check photosynthesis by restricting the entrance of carbondioxide as well as by acting as a direct limiting fact. Similarly high light intensities often cause increased rates of transpiration .

In directly this causes a reduced water content of the leaf cells, which may inturn cause a diminished rate of photosynthesis. A high light intensity also has the usual effect of raising leaf temperatures some what above the prevailing temperatures of the surrounding atmosphere and may thus influence photosynthesis indirectly by its effect on thermal conditions within leaves. Very high light intensities also have a destructive effect upon chlorophyll and other cell constituents as mentioned in the preceding paragraph.

Effects of Different Light Qualities upon Photosynthesis

Hoover (1937) investigated the effect of different wave lengths of radiation upon the rate of photosynthesis of wheat plants. By the use of suitable filters he was able to irradiate the plants with narrow spectral bands. All measurements were made with equal but low intensities of radiation incident upon the plants and under such conditions that no other factor was limiting for photosynthesis. The results of his investigation indicate the occurrence of maximum photosynthesis at a wave length of 655 mμ in the red, and a secondary maximum of 440 mμ in the blue. It is worthy of note that this curve does not show a very close agreement with absorption spectrum of chlorophyll.

The results of all investigators on the relation between light quality and photosynthesis do not agree with those just described. Gabrielsen (1948), for example, found in several species, including wheat, that the highest rates of photosynthesis occurred in orange-red light, the next highest in green-yellow light, and the lowest in violet-blue light. With *Ulva lactuca* (a marine alga), however, results were obtained similar to those of Hoover for wheat.

It is not certain whether the differences in the effect of light quality on photosynthesis from one species to another which appear to exist are real or not. It is possible that the relation between light quality and photosynthesis is not the same in all kinds of plants, and may even be different in two plants of the same species which have developed under different conditions. On the other hand, the technical difficulties in this type of experimentation are considerable, and it is not clear just how much of the available data has been obtained under sufficiently critical conditions. It is also possible, therefore, that the diverse results which have been recorded are more apparent than real, and are largely they outcome of the different experimental techniques which have been used. There are a number of conditions under which plants growing in their natural habitats are exposed more or less continuously to light of a different quality from that of full sunlight at the earth's surface. On cloudy days, for example, the intensity of light is not only less than on clear days, but it is proportionately richer in blue and green wave lengths.

Light which has been filtered through the crown of a tree is usually proportionately richer in green rays than direct sunlight because of the greater proportionate absorption by leaves in the red and blue portions of the spectrum. This effect upon light quality is most marked in hardwood forests in which the tree crowns form an almost continuous canopy. The herbs, shrubs, and smaller trees growing in such forests are subjected to light which is not only of much lower intensity than full sunlight, but is also different in quality from the light impinging upon the forest canopy.

In habitats of submerged aquaties both the intensity and quality of the light are usually very different from the intensity and quality of the sunlight at the earth's surface. Pure water absorbs radiations in the red-orange portion of the spectrum much more effectively than in the blue-green region. While the absorption coefficients or natural waters for various wave dissolved or dispersed in the water, in general shorter wave lengths penetrate to greater depths than longer wave, lengths. Hence with increasing depth in either fresh or ocean water not only is the intensity of the light reduced, but its quality is greatly modified. Aquatic plants growing at a depth of 20 much richer in blue-green rays, although of lower intensity, than those at a depth of 1 meter.

Alpine plants are also exposed to light of different composition than species at lower attitudes. The atmosphere absorbs the shorter wavelengths of the sun's radiation more effectively than the longer once. Because of the shorter column of atmosphere through which it passes, sunlight at high altitudes is also relatively richer in the shorter wave lengths of visible radiation and the ultraviolet.

Effects of Duration of the Light Period upon Photosynthesis

In general a plants will accomplish more photosynthesis in the course of a day if exposed to illumination of favourable intensity of ten or twelve hours than if suitable light conditions prevail for only four or five hours. In arctic regions photosynthesis may occur continuously through out the 24-hr day of the summer months although the rate may vary cyclically throughout the day. A considerable capacity for sustained photosynthesis under continuous light is indicated by the results of Bohning (1949). Leaves on young apple trees, exposed to a continuous illumination of about 3200 foot candles at 25°C, and the usual atmosphere concentration of carbon dioxide, were shown to photosynthesize at an undiminished rate of periods of at least 18 days.

Temperature Effects on Photosynthesis

The measurement of the effect of temperature upon photosynthesis in terrestrial plants is complicated by the fact that the leaf temperatures of such plants are seldom the same as atmospheric temperatures. Whenever leaves are exposed to direct illumination, which is almost invariably the situation when photosynthesis is occurring at a rapid rate, their temperatures exceed those of the surrounding atmosphere. It is difficult, therefore, if not impossible, to maintain the temperature of the leaves of land plants at a desired value while they are exposed to light of any considerable intensity. Evaluation of the effect of the temperature of leaves upon photosynthesis is possible only if the actual leaf temperature is measured. For this reason many of the more critical studies of the effect to temperature upon photosynthesis have been made with submerged water plants, in which a close thermal equilibrium is maintained between the plant body and the surrounding water.

Temperature Limits of Photosynthesis

Photosynthesis can take place over a wide range of temperatures. It has been reported to occur in some species of conifers at temperatures as low as – 35°C. and in some kinds of lichens at – 20°C. Freeland (1944) has shown that the photosynthetic rate may exceed the rate of respiration in several species of conifers at temperatures as low as – 6°C. Tropical plants cannot carry on photosynthesis at temperature as low as those at which many temperate zone plants do. In most tropical species photosynthesis apparently will not occur at temperatures below about 5°C. At the other end of the temperature range for photosynthesis stand the species of algae indigenous to hot springs which can survive 75°C. and probably carry on photosynthesis at temperatures close to this value. Many semi desert and tropical species can withstand air temperatures of 55°C. and probably photosynthesize at temperatures not far below this. In most plants for temperature regions the range of temperatures within which photosynthesis occurs at a relatively rapid a rate is about 10°-35°C.

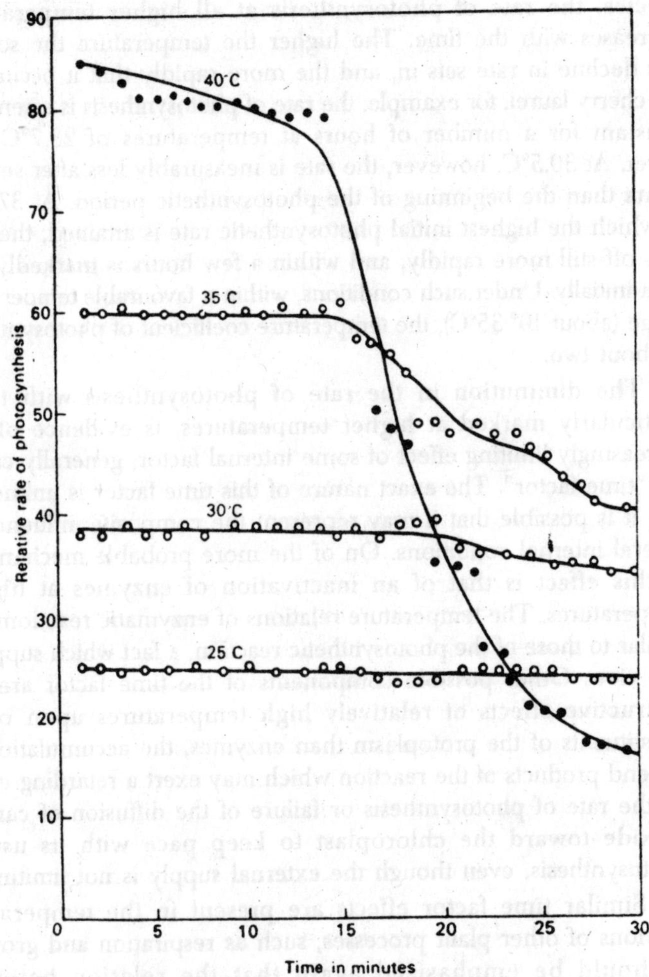

Fig. 5.8. Effect of temperature on the photosynthesis.

The Effect of Temperature on the Rate of Photosynthesis

If neither carbon dioxide, not light, nor any other factor is limiting, the rate of photosynthesis increases with rise in temperature up to a point which varies somewhat from one kind of a plant to another. Further increase in temperature results in a rapid decline in the rate of photosynthesis, resulting primarily from injurious effects of higher temperatures on the protoplasm. Furthermore, as first clearly shown by Mattaei (1905) for cherry laurel (*Prunus laurocerasus*) but since confirmed by other investigators for other species, the rate of photosynthesis at all higher temperatures decreases with the time. The higher the temperature the sooner this decline in rate sets in, and the more rapidly that it occurs. In the cherry laurel, for example, the rate of photosynthesis is essentially constant for a number of hours at temperatures of 23.7°C and lower. At 30.5°C, however, the rate is measurably less after several hours than the beginning of the photosynthetic period. At 37.5°C, at which the highest initial photosynthetic rate is attained, the rate falls off still more rapidly, and within a few hours is markedly less than initially. Under such conditions, within a favourable temperature range (about 10°-35°C), the temperature coefficient of photosynthesis is about two.

The diminution in the rate of photosynthesis with time, particularly marked at higher temperatures, is evidence of the increasingly limiting effect of some internal factor, generally called the "time factor". The exact nature of this time factor is unknown, and it is possible that it may represent the composite influence of several internal conditions. On of the more probable mechanisms of this effect is that of an inactivation of enzymes at higher temperatures. The temperature relations of enzymatic reactions are similar to those of the photosynthetic reaction, a fact which supports this view. Other possible components of the time factor are the destructive effects of relatively high temperatures upon other constituents of the protoplasm than enzymes, the accumulation of the end products of the reaction which may exert a retarding effect on the rate of photosynthesis or failure of the diffusion of carbon dioxide toward the chloroplast to keep pace with its use in photosynthesis, even though the external supply is not limiting.

Similar time factor effects are present in the temperature relations of other plant processes, such as respiration and growth. It should be emphasized again that the relation between photosynthesis and temperature described above holds only when no other environmental factor is limiting. Such experiments on the overall temperature characteristics of the process must therefore be conducted with the plants exposed to an atmosphere in which the carbondioxide concentration is considerably higher than the usual atmospheric concentration of 0.03 per cent. In nature the maximum rate of photosynthesis which might be achieved under the prevailing temperature is often not realized because of the limiting effect of some other factor.

Temperature, within the ordinary physiological range for plants, has little effect on the rate of photosynthesis of plants growing in deep shade or of plants exposed to the low light intensities of cloudy days. Under such conditions lights the limiting factor. Similar, on warm days, well-watered plants exposed to bright light often do not photosynthesize at the maximum rate possible under the prevailing temperature because carbon dioxide at atmospheric concentration is the limiting factor. Thomas and Hill (1949), for example, found that temperature had very little influence on the rate of photosynthesis of alfalfa under field conditions within a range of 16.4° to 29°C. In other words the temperature co-efficient of the process under such conditions is approximately one.

The Role of Water in Photosynthesis

Less than 1 percent of the water absorbed by a plant is used in photosynthesis. It therefore seems probable that the indirect effects of the water factor upon photosynthesis are more pronounced than its direct effect. In other words, deficiency of water as a raw material is rarely if ever a limiting factor in photosynthesis. Nevertheless a reduction in the water content of leaves usually results in a decrease in the rate of photosynthesis as is illustrated

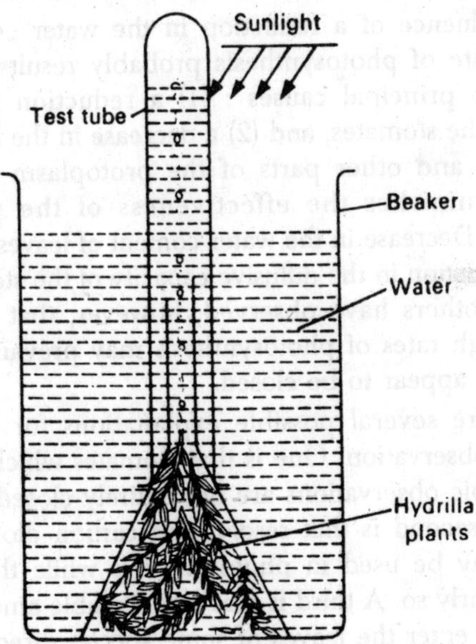

Fig. 5.9. Liberation of oxygen in photosynthesis by an aquatic plant.

by the results of Schneider and Childers (1941). These investigators studies the effects of withholding water upon the rate of apparent photosynthesis in apple trees. When the soil was allowed to dry out gradually, starting from the field capacity, reduction in the rate of apparent photosynthesis (and transpiration) became evident within a few days at an air temperature of 26.7°C, and was pronounced before visible wilting took place.

When the soil water content had fallen to the permanent wilting percentage, and the leaves were distinctly wilted, apparent photosynthesis was 87 percent less than the initial rate. Upon watering the soil the leaves regained turgidity within a few hours, but the original rates of apparent photosynthesis were not reattained until 2-7 days had passed. An important implication of this last observation is that the retarding effects of drought on photosynthesis linger for some time after water has again become available in the soil. Submersion of roots also has a pronounced retarding effect on apparent photosynthesis (and transpiration) in apple, usually becoming evident within 2-7 days after flooding the soil with water. With continued water-logging of the soil around the roots, rates of apparent photosynthesis often become immeasurable. This effect is doubtless largely the result of a retardation in water absorption.

The influence of a reduction in the water content of leaves upon the rate of photosynthesis probably results from either or both of two principal causes : (1) a reduction in the diffusive capacity of the stomates, and (2) a decrease in the hydration of the chloroplasts and other parts of the protoplasm which in some manner diminishes the effectiveness of the photosynthetic mechanism. Decrease in the water content of leaves unquestionably causes a reduction in the diffusive capacity of the stomates. Mitchell (1936) and others have observed, however, that in some plants relatively high rates of photosynthesis may prevail at times when the stomates appear to be closed.

There are several possible explanations for this apparently anomalous observation. One is that stomates which appear closed to microscopic observations are not entirely closed to the passage of gases. A second is that reservoired carbon dioxide within the leaf cells may be used in photosynthesis while the stomates are closed or nearly so. A third is that considerable amounts of carbon dioxide may enter the leaves of some species directly through the epidermis as described earlier. The evidence in favour of the view that the decreased hydration of the protoplasm which accompanies a reduction in cell turgidity causes a decreased rate of photosynthesis comes chiefly from experiments with water plants.

Walter (1929), for example, has studied the effect upon their rate of photosynthesis of immersing plants of waterweed (*Anacharis canadensis*) in sucrose solutions of various concentrations. The greater the concentration of the sucrose solution the less the turgidity of the cells, the less the hydration of protoplasm and the slower the rate of photosynthesis. In one experiment the rate of photosynthesis was appreciably retarded by immersion of the plants in an 0.3 molal solution and almost entirely stopped in an 0.7 molal solution of sucrose with an osmotic pressure of about 18atm. Plasmolysis of the cells occurred at a solution concentration of between 0.3 and 0.4 molal.

The rate of respiration, on the other hand, was practically unaffected by sucrose solutions at any concentration up to and including 1.0 molal. Since the water weed is a submerged aquatic with thin leaves which bear no stomates, any reduction in the rate of photosynthesis resulting from a diminution in cell turgor most likely results from direct effects upon the hydration of the protoplasm of the photosynthesizing cells. In general, it appears that the rate of photosynthesis is less affected by reduction in leaf water content than the rate of transpiration. This is indicated by the results of Heinicke and Childers (1936) who determined the average rates of photosynthesis and transpiration over a one-week period in two apple trees, one of which was well watered while the other was growing in soil which was gradually drying out. The rate of photosynthesis of the plant in the dry soil was about half as great as in the plant which was watered. The rate of transpiration, however, was only about one-fourth as great in the former plant as in the latter.

Effect of Oxygen Concentration on Photosynthesis

The photosynthetic organs of terrestrial plants are seldom exposed to oxygen concentrations which deviate appreciably from the usual atmospheric concentration of about 21 percent. Hence studies of the effect of oxygen concentration on the rate of photosynthesis are principally of theoretical interest. In at least some, and probably all plants, increase in the oxygen concentration results in a decrease in the rate of photosynthesis, and the normal atmospheric concentration is sufficiently high to induce a slower photosynthetic rate than obtains at lower oxygen concentrations. For example, MeAlister and Myers (1940) showed that the rate of photosynthesis in young wheat plants was about 30-50 percent higher in 0.5 percent oxygen than in 20 percent oxygen at a high light intensity and atmospheric carbon dioxide concentration. This effect of oxygen is not merely a decrease in apparent photosynthesis resulting from and enhanced respiration rate at the higher oxygen concentration but is on the process of photosynthesis *per se.* Neither is this effect one of photo-oxidation offsetting photosynthesis, because the magnitude of the retardation in photosynthesis under relatively high concentrations of oxygen is too great to be accounted for in this way. The only satisfactory explanation appears to be that oxygen actually exerts a direct inhibitory effect on photosynthesis, and the greater the concentration of oxygen the greater this effect.

Effects of Certain Chemical Compounds on Photosynthesis

Many different kinds of chemical compounds, upon absorption by plant cells, have effect–direct or indirect–upon the rate of photosynthesis. Especially noteworthy are the effects of certain substances which markedly influence the rate of photosynthesis when present only in minute quantities. Among these are hydrocyanic acid, hydroxylamin, hydrogen sulphide, and certain compounds containing the iodoacetyl radical. Hydrocyanic acid, for example, has a pronounced retarding effect on the photosynthesis of *chlorella* at a concentration of only 4×10^{-5} molar. All of these substances appear to act as enzyme inhibitors and each of them appears to affect one specific enzyme system of the several which operate in the mechanism of photosynthesis. This fact has been used to advantage by investigators of the chemical dynamics of photosynthesis, who have thereby been enabled to inhibit the process at different stages, and thus to differentiate more clearly among the several partial reactions of photosynthesis. For example, the enzyme system which is sensitive to hydroxylamine participates only in the oxygen-releasing stage of photosynthesis.

Narcotics, of which chloroform, ether, and the urethans are examples, also exert retarding or inhibitory effects on photosynthesis when present in very low concentrations, unless was the concentration is too high, in which case death of the cells may result, the effect of such substances is reversible. Narcotic inhibition of photosynthesis does not appear to result from effects on specific enzymes, but from some more general influence, which is believed to be retated to their surface activity.

The Effects of Internal Factors on the Rate of Photosynthesis

In addition to the environmental factors which influence photosynthesis, the rate of the process is also influenced by certain factors within the plant. In general, because of the experimental difficulties encountered in their investigation, the effects of such internal factors ion photosynthesis are less well understood than those of external factors. Some discussion of several of these factors is warranted, however, by the present state of our knowledge.

Chlorophyll Content

In their extensive study of the relationship between chlorophyll content a:.d photosynthesis Willstatter and Stoll (1918) devised the *photosynthetic number* ("Assimilationszahl") as an index to this relation. The photosynthetic number is the number of grams of carbon dioxide absorbed a per hour per gram of chlorophyll. These two investigators studied the relation between chlorophyll content and photosynthesis in green-leaves and yellow-leaves varieties of the same species. In this experiment the leaves were exposed to strong light in an atmosphere of 5 percent carbon dioxide at 25°C. The rate of photosynthesis in green leaves varieties is not much in excess of that in yellow-leaves varieties of the same species, and, when expressed in terms of the photosynthetic number, the yellow-leaves varieties are much more efficient per unit of chlorophyll present.

Other investigations by the same workers also seem to point to the conclusion that there is no proportional relationship between chlorophyll content and photosynthesis in the leaves of vascular plants. In other words, it appears that the chlorophyll content of the leaves is seldom the limiting factor in photosynthesis in such species even when all external conditions are favourable for the process.

Hydration of Protoplasm

That the hydration of the protoplasm is an important internal factor influencing photosynthesis has already been shown in the discussion of the water factor in photosynthesis.

Leaf Anatomy

The rate of photosynthesis in any leaf is partly conditioned by the anatomy of that leaf. The size and distribution of the intercellular spaces, the relative proportions and distribution of palisade and spongy layer, the size, position, and structure of the stomata's, the thickness of cuticular and epidermal later, the amount and position of sclerenchyma, proportions and distribution of non-green mesophyll tissues, and the size, distribution and efficiency of the vascular system, all influence the rate of photosynthesis.

Protoplasmic Factors

Evidence from various types of experiments, some of which has been presented on the foregoing pates, indicate conclusively that certain other conditions resident in the protoplasm of plants cells, other than chlorophyll content and hydration, influence that rate of photosynthesis. The most important of such factors is undoubtedly the enzyme complement of the protoplasmic system. That a number of enzymes play a part in photosynthesis is beyond the question. There is also little doubt that the proportions and quantities of these enzymes may varying different kinds of photosynthetic cells, or in the same cell under different conditions, but until more is known of the physicochemical nature of the photosynthetic catalysts on discussion of them beyond these generalized statements is warranted.

Accumulation of the End Products of Photosynthesis

During photosynthesis the carbohydrates synthesized in the process, or in immediately following secondary reactions, accumulate in the photosynthesizing cells more rapidly than they are translocated toward other tissues. Under at least some conditions accumulation of carbohydrates appears to exert a retarding effect upon photosynthesis. The results of Boysen-Jensen and Muller (1929), Kjar (1937), and others, however, appear to indicate that there is no consistent relationship between the concentration of carbohydrates in the leaf cells of intact plants and the rate of photosynthesis.

Daily Variations in the Rate of Photosynthesis

Thomas and Hill (1937) measured the daily variation in rates of apparent photosynthesis for small plots of alfalfa under conditions approximating a "standard day". Well-watered plots of alfalfa 6 feet square were enclosed in transparent celluloid cabinets and air circulated through each cabinet at ranging up to several hundred cubic feet per minute. The net consumption of carbon dioxide by the plants was ascertained by measuring the difference in the concentration of this gas in the inflowing and outflowing streams of air.

In general, as with the apple tree previously described, the rate of photosynthesis under these conditions showed a close correlation with the light intensity. Innumerable other types of daily periodicities of photosynthesis are possible. The pattern of any such periodicity depends in part on the kind of plant and on the unit of plant material for which photosynthesis is measured–whether a single leaf, a small plant, a large plants, such as a tree, or a plot of vegetation, It will also obviously vary with the daily cycle of environmental conditions which not only differs in a general way from one climatic center to another, but also shows seasonal and daily variations within any climatic region.

6

EFFECT OF HEAT STRESS ON PHOTOSYNTHESIS

Environmental stresses, such as those resulting from rapid changes in nutrition, water supply, anaerobiosis and temperature limit the productivity and adaptability of crops. Heat stress affects crop productivity by influencing almost all the physiological and biochemical processes from seed germination to grain development. If heat stress occurs in the life cycle of a crop plant, it is important that it should possess a certain degree of heat tolerance to survive the stress period. Adaptation to high temperature has been reported for many wild plants. Though evidently based on some physiological parameters, heat-tolerant genotypes of some commercial crops have been identified the exact molecular mechanism of heat tolerance has not yet been understood. A thermoinducible system of heat-shock protection is known to be present in several organisms including plants. In these cases, certain genes are induced which result in the synthesis of heat-shock polypeptides. None of these polypeptides have so far been identified and, moreover, protection is provided for a very short period. However, adaptation of crop plants to high temperature has been reported to be correlated with photosynthesis.

High temperature causes membrane disruption, which alters solute movement, photosynthesis and respiration. Sheoran *et al.* have developed a wheat mutant, WH147M, which has been reported to possess high thermal requirement. Seedlings of WH147M were found to have a relatively better heat-inducible thermoprotection mechanism. Results reported here show that the thermotolerant WH147M plants possess relatively more thermostable photosynthetic and respiratory activities.

MATERIAL AND METHODS

Plant Material

Seeds of wheat (*Triticium aestivum* L.) *cv.* WH147 and its thermo-tolerant mutant, WH147M, were obtained from the Department of Plant Breeding of this University. The crop was raised in pots in the *rabi* season, using standard agronomic practices.

Temperature Treatment

The plants, at the tillering stage, were subjected to heat stress by placing pots in a BOD incubator at the desired temperature. Heat stress, during ion-leakage determination, was effected by placing leaf discs on a temperature-controlled shaker water-bath.

Ion leakage

Leaves were washed and cut into small discs, using a punch. Ten discs were placed in a test tube containing 10 ml deionised water. The test tube was kept for heat-stress in a shaker water-bath at the desired temperature. The sample was cooled to room temperature and the conductivity of the liquid was measured. The same sample was then placed in a boiling- water bath for 10 min. The water loss during boiling was replenished, the sample was shaken for 30 min, cooled and conductivity was measured. The values were expressed as percentage of the total conductivity.

Photosynthesis

The photosynthetic rate was measured with an Infra Red Gas Analyser, model ADC type 225/2K. The measurements were made at 11 a.m at an average light intensity of 1000 µ E m^{-2} S^{-1}. The leaves were cut under water and placed in a 13 × 8 × 8 cm airtight Perspex chamber. Leaf-area was measured using a Leaf-Area Meter, model L1-3000, and the rate of photosynthesis expressed as mg $CO_2/dm^2/h$.

Oxygen Uptake Activity

Oxygen uptake activity was measured polarographically, using a oxygen electrode. Leaf discs were placed in the monitor cuvette containing 0.1 M phosphate buffer pH. 7.5. Oxygen uptake activity was expressed as µl O_2 consumed h/g fresh weight.

TTC Reduction Activity

Small discs punched from the leaves of temperature-stressed plants were weighed and kept in 0.05 per cent Triton X-100. The discs were incubated at 30°C, 15 *h*, in the dark and then formazan was extracted in 95 percent ethanol in a boiling-water bath for 5 min. Absorbance of the colour was measured at 530 nm and the results expressed as absorbance units/g leaf.

RESULTS

Ion-leakage

Ion leakage from leaves was estimated by measuring the electrical conductance of leachates from leaf discs. The results, presented in Table 6.1 show that temperature treatment at 50° and 55°C, for varying periods, resulted in more ion-leakage from WH147M discs. WH147 leaf discs as compared to that from WH 147M discs. A 55°C treatment for 20 min resulted in 100 per cent leakage (total leakgage determined by a 100°C treatment for 10 min) in WH147 leaf discs, while in WH147M it was 76 per cent.

Table 6.1. Temperature Stress-Effected Ion-leakage from WH147 and WH147M Leaves.

Treatment °C, min	*Ion-leakage (per cent conductivity)*	
	WH147M	*WH147*
50°C		
45 min	42.8	45.4
90 min	71.4	81.8
55°C		
5 min	19.0	31.3
10 min	38.0	62.5
20 min	76.1	1.00

Tetrazolium Reduction Activity

Results presented in Table 6.2 show that temperature stress of 45°C for 40 min did not appreciably affect the tetrozolium reduction activity of leaf discs of the parent, while an increase was observed in the mutant. When stress treatment of 50°C was given the decrease in tetrazolium reduction activity ws appreciable in WH 147, while it increased in WH147M leaf discs. The 40 min stress treatment resulted in nearly 50 percent reduction in the WH147, whereas in WH147M there was an increase of 28 per cent.

Table 6.2. Tetrazolium Reduction Activity of Temperature-stressed WH147 and WH147M leaves.

Treatment °C, *min*	*Tetrazolium reduction activity (OD units g^{-1} leaf)*	
	WH147M	*WH147*
45°C		
20 min	8.9 (+42)	9.1 (+2)
40 min	10.9 (+73)	8.8 (–13)
50°C		
20 min	10.1 (+66)	5.3 (–39)
40 min	8.1 (+28)	4.2 (–52)

Oxygen Uptake Activity

Temperature stress resulted in a higher reduction in oxygen uptake activity of WH 147 leaves as compared to WH147M leaves (Table 6.3). Treatment at 50°C for 45min resulted in 72 per cent reduction in oxygen uptake by WH147 leaves, whereas the corresponding decrease in WH147M leaves was only 36 per cent.

Table 6.3. Oxygen Uptake Activity of Temperature stressed WH147 and WH147M leaves.

Treatment °C, *min*	*Oxygen uptake (ml O_2 min^{-1} g^{-1} leaf)*	
	WH147M	*WH147*
50°C		
15 min	294 (–15)	238 (–60)
30 min	203 (–31)	207 (–66)
45 min	220 (–36)	167 (–72)

Photosynthetic Activity

Table 6.4 shows the results of photosynthetic activity of WH147M and WH147 temperature-stressed wheat leaves. Reduction in photosynthetic activity due to temperature stress was relatively more marked in WH147 leaves as compared to WH147M leaves. Stress of 45°C for 60 min resulted in 82 and 33 percent reduction in WH147 and WH147M respectively. At 50°C treatment for 1*h*, the leaves of WH147 completed lost photosynthetic activity where as the loss was only 30 per cent in WH147M leaves.

Table 6.4. Temperature Stress-Effected Ion-leakage from WH147 and WH147M Leaves.

Treatment *°C, min*	*Photosynthetic activity* *(mg CO_2 dm^{-2} h^{-1})*	
	WH147M	*WH147*
45°C		
30 min	21.4 (–63)	12.5 (–40)
60 min	22.8 (–33)	5.6 (–82)
50°C		
30 min	21.3 (–37)	1.7 (–95)
60 min	23.6 (–30)	

Discussion

Thermo-insensitive wheat mutant WH147M is a product of gamma irradiated parent WH147. Grain yield of the mutant is highest when it is sown in the high temperature day of September (mean maximum temperature, 35.5°C, minimum temperature 21.0°C), as compared to the normal sowing time of November (mean maximum temperature 29.1°C, minimum temperature 10.6°C) for the parent. Thus, the mutant possesses some thermotolerant developmental mechanism which facilitates normal crop production even under adverse climatic conditions. Under different stress conditions, alterations in gene expression takes place in crop plants. Heat-stock polypeptides are known to be synthesized in response to heat stress in soyabean maize and wheat.

The mechanism by which heat-shock polypeptides provide protection against heat stress in plants is not understood. However, their intracellular localization, particularly of hsp 18, with plasma membrane suggests that they protect membrane destruction. In WH147M the membrane damage caused by simulated temperature stress, as indicated by ion-leakage, was less as compared to that in WH147. Electrolyte leakage has been reported to be a good index of heat-stress tolerance in pear tomato, barley and onion bulbs soyabean and sorghum.

The maintenance of membrane integrity under temperature stress is also indicated by an active endoplasmic reticulum. Belanger *et al.* reported the destabilisation of mRNA and disruption of rough endoplasmic reticulum at high temperature is barley aleurone. In the leaves of WH147M, incorporation of H-leucine increased at high temperature, while in parent WH147 there was a reduction in H-leucine in corporation. The respiratory activity of WH147M, as indicated by tetrazolium dye reduction and oxygen uptake activity also showed a thermotolerant character.

Tetrazolium reduction activity is a suitable selection test for screening of heat-tolerant genotypes. In WH147M leaves, heat stress increased the tetrazolium reduction activity, while the oxygen consumption decreased. This perhaps was due to a reduction in respiratory activity without any change in the energy charge, as has been reported in pea seedlings subjected to high temperature. The photosynthetic apparatus of WH147M was also relatively more thermotolerant than that of Wh147.

Heat stress inhibits photosynthesis by affecting chlorophyll accumulation, protein synthesis and photosystem activities. Synthesis of chloroplast proteins, particularly ribulose 1, 5 diphosphate carboxylase of wheat is already known. Though the molecular mechanism of heat tolerance in WH147M is not understood, this mutant has opened up a possibility of wheat cultivation under wider agroclimtic conditions, particularly under early sown and rainfed conditions, which though desired, was hitherto not feasible.

7

GENETIC CONTROL OF PHOTOSYNTHESIS

The chloroplast, a miniature green laboratory of a plant cell, possesses a number of surprising properties. A unique one is its capacity to convert light energy into chemical energy to accomplish the most important biological process of primary solar energy storage on the earth. Photophysical, photochemical, and biochemical aspects of photosynthesis have been elucidated well enough in several reviews. At present, genetic aspects of photosynthesis attract the attention of many scientists–plant physiologists and biochemists, molecular biologists and biophysicists, geneticists and plant breeders. This interest is natural because the problem is very important and complex and its solution requires and combined physiological and genetic approach. The subject involves two basic biological processes: (a) expression of genetic information, and (b) regulation of photosynthesis.

Investigation of the genetic nature of photosynthesis will enable us to understand the molecular mechanisms of self organization and autoreproduction of chloroplasts, their inheritance and variability, the processes of plastid biogenesis, and the formation of its functional activity. On the other hand, the study of genetic determination of chloroplast structure and function will make it possible to enrich breeding search with new methods and tests and to improve photosynthetic plant productivity. Regulation of photosynthetic efficiency through inheritance is an important means of enhancing of enhancing crop yield.

Genetics of Photosynthesis

The genetic material of eukaryotic organisms is localized not only in the nucleus, which is the main store of hereditary information, but it is dispersed throughout the cell is sub cellular structures: mitochondria, kinetosomes, chloroplasts. The plant genotype as a complex of all cell genetic factors is subdivided into chromosomal–nuclear genome–and extra chromosomal–plasmone, comprising the cytoplasmic determinants, mitochondria and chloroplasts. Interaction of nuclear and cytoplasmic genetic systems with one another and with the environment determines both the course and the dynamics of cell differentiation, plant growth, and development.

Interaction of plant cell genetic factors has a very complex but rigid character. The nucleus as an integrating centre of genetic information is responsible for the strategy of formation of constant organism- specific properties. Cytoplasmic genes and organelles determine the cell life tactics, providing its greater flexibility and adaptation to environmental conditions.

Cell organelles are closely interrelated in their development and functional activity. Their cooperative interaction provides the integration of information, energy, and substance flow in cell activity. A living cell is not a sum of autonomous or semi-autonomous organelles, but a unique and highly organized dynamic system attaining new properties differing from those of single organoids due to "concert" interaction of subcellular structures.

Regulation of photosynthesis requires close cooperation the discrete genetic systems of a plant cell. The chloroplast genome is not large enough to provide all the diverse reactions of this complex process, so a considerable part of the genetic information necessary for chloroplast development and activity is localized in the nucleus. In order to integrate photosynthetic activity in cell metabolism, it is possible that in the course of evolution chloroplasts acquired a double genetic subordination–to the nuclear genome and to their own semiautonomous one.

Hereditary properties of chloroplasts were discovered long ago, whereas the study of the genetic nature of photosynthesis as a whole began only recently. We would say that genetics of photosynthesis is a new trend in plant physiological genetics. It can be defined as an interaction between the balanced plant cell genomes which control self-organization and reproduction of chloroplasts, their hereditary continuity and variability, carbon pathways, and photosynthetic potential.

History

Non-Mendelian plastid inheritance was discovered by Correns in spontaneous variegated mutants of *Mirabillis jalapa.* Crossing normal and variegated forms, he showed that, independently of the pollen type, the pigmentation of progeny is determined exclusively by the egg cell. Proceeding from this fact, Correns postulated the principle of maternal or cytoplasmic inheritance which he called *Status albamaculatus.* Baur made an important contribution to the understanding of the nature of cytoplasmic inheritance and attempted to give it a scientific explanation. He showed that in *Pelargonium zonale* both parents contributed to plastid inheritance. This principle was called a biparental cytoplasmic inheritance, *Status Paraalbomaculatus.*

Baur was the first to assume that chloroplasts possess their own hereditary apparatus responsible for their inheritance and genetic continuity. It is plastid genome mutation that causes the appearance of variegated plants. The intensive study of the mechanism of cytoplasmic inheritance is connected with the development of ideas and methods of molecular biology and molecular genetics. At the beginning of the 1960s a specific DNA was found in chloroplasts. Its molecule differed significantly from that of nuclear DNA in size, nucleotide content, and other physicochemical properties. At the same time, plastid ribosomes of the prokaryotic type were found which differed from the cytoplasmic ones in their sedimentation properties. These findings spurred the rapid development of investigations into genetics of chloroplasts and their transcription translation system (TTS).

Using gradient ultracentrifugation techniques in the presence of acridine dyes, as well as electron microscopy, autoradiography, and molecular hybridization, many advances have been made in the study of the structure and information capacity of the chloroplast, the mechanisms, of its self-replication and transcription. It has been established that chloroplast DNA is associated with the lamellar system and has different configurations : linear, circular, loop shaped, and superhelical. Plastid genome replication has a semiconservative character.

Optical registration of DNA renaturation made possible determination of the chloroplast genome size in different plants.

An organelle DNA ranges from 1 to 2 × 10^8 daltons, which corresponds to 10^6 pairs of nucleotides. The DNA content in one chloroplast is 0.1 to 1.2 × 14^{-14}g, depending on plant species. These data make it possible to calculate the size of the chloroplast genome. The number of DNA copies in an individual chloroplast is 20 to 60.

Chloroplasts contain all the components of the protein synthesizing system (PSS): ribosomes, RNA-polymerase, mRNA, tRNA, aminoacyl-tRNA synthetases, and other translation factors. Etioplasts and chloroplasts isolated from green leaves are able to synthesize nucleic acids and structural and functional proteins. These data served as the basis for development of the concept of genetic and biochemical autonomy of chlorpolasts. An opinion was even expressed that chloroplasts have an independent genetic apparatus and resemble "a cell within a cell." This was an additional argument in favour of the symbiogenetic hypothesis of its origin.

Division of chloroplasts in vitro found by Ridley and Leech speaks in favour of their genetic and biochemical autonomy and confirms the possibility of obtaining self-reproducing cultures of isolated chloroplasts. However, despite numerous attempts, nobody has yet succeeded in obtaining a self-reproducing and functionally active chloroplast culture. In experiments by Wellburn and Wellburn with etioplasts isolated from oats, disorganization of outer and inner membranes was observed even after 2 hr of illumination, and after 9 hr the internal structures were completely destroyed and eliminated. Studying biogenesis of isolated maize etioplasts. Leech came to the conclusion that chloroplast development is the result of collaboration of the whole cell, and "longliving" cultures of etioplasts and proplastids can be obtained by addition of a great number of cofactors and even other organelles to the suspension.

Differentiation of plastids depends on interaction of both chloroplast and nuclear genome. Indeed, possessing unique genetic information and TTS, the chloroplast is not an autonomous organelle, "a cell within a cell," because its genetic and biochemical function is semiautonomous. The size of the plastid genome is not sufficient to code for the formation of the great amount of structural and functional proteins which are necessary for proliferation of membrane systems and for carrying out diverse light and dark reactions of photosynthesis. Numerous available experimental data testify to the essential role of the nucleus in chloroplast biogenesis.

Using different physiological and chemical agents, we induced a wide spectrum of higher plant mutations blocking different stages of biosynthesis of chlorophyll, components of the photosynthetic electron transport chain (ETC), and Calvin cycle enzymes. Sixty nonallelic genes affecting biogenesis of plastid pigments were fcund in the model object *Arbidopsis thaliana* Heynh. Experimental mutagenesis combined with investigations using specific inhibitors of TTS enabled us to dissect the genetic regulation of synthesis of lamellar proteins, plastic enzymes pigments, and formation of chloroplast photosystems. With these investigations as a basis, we have suggested a conception of nuclear-plastid cooperative interaction in the development of chloroplast structure and function.

Structure and Organisation of Genes for Proteins of the Photosynthetic Membranes

Genes for the photosynthetic apparatus proteins are located both in the chloroplast and nuclear genomes. The first indication of the distribution of these genes in two compartments came from early experiments on hybridization with various genotypes and mutants which indicated that whereas with respect to certain characters inheritance followed the Mendelian pattern in respect of others it did not. In the early seventies, experiments began employing compartment-specific inhibitors of protein synthesis. These experiments were followed by *in vitro* protein synthesis from mRNA isolated from cells–by separating poly A mRNA from other RNAs, one could further confirm whether the nucleus or the chloroplast DNA coded for a particular product. In the late seventies, chloroplast genomes (cpDNAs) were characterised from several plants e.g., spinach and maize. Although some variations do exist, most cp DNAs have turned out to be circular molecules of 120-180 kbp and containing a pair of 10-25 kbp inverted repeats, which divide the remaining cpDNA into one small single copy region.

Finally, with pure cpDNA available from several plants as also various restriction endonucleases and recombinant DNA technology, the stage was set for detailed characterisation of the chloroplast genome and for localisation of the genes encoded by the plastome on cloned cpDNA fragments. Genes were assigned to various proteins by using the immunological approach, following one of the two preparative strategies for obtaining the proteins: (i) the DNA fragments were cloned and transcription coupled with translation employing cell-free extracts of *E. coli* itself, (ii) cp DNA fragments were used to hybrid select specific mRNAs which were then translated *in vitro* using the reticulocyte system. Many cloned cpDNA fragments have now been sequenced and open reading frames (ORFs) for various polypeptides identified.

The deduced sequences have also been compared to partially or completely sequenced polypeptides. In 1986, the complete nucleotide sequences of cpDNAs from a higher plant, tobacco (*Nicotiana tobacum*), and a liverwort (*Marchantia polymorpha*) were reported. Several ORFs have already been identified and efforts are being made to find products for those ORFs whose function is not yet known. Rapid progress has similarly been made in respect of genes residing in the nucleus. in the early eighties, cDNA libraries from a number of plants were made and clones encoding chlorophyll *a*/*b* binding polypeptides of PSII were identified. The cDNA clones were then used to isolate genomic clones. More recently, a cDNA library of spinach has been made in expression vectors and clones, specific to several proteins of the photosynthetic membranes, have been isolated by using antisera.

The cDNA clones have been employed to isolate and characterise genomic clones. Interestingly, all these genes encode precursor polypeptides with amino-terminal transit peptides which help to target these polypeptides into the chloroplast and which are removed during the transport across the chloroplast or thylakoid membranes. Some details about the organisation and characteristics of the genes encoding polypeptides of various supra-molecular complexes and other associated proteins of the photosynthetic membranes are given in Table 7.1. A general summary follows.

Photosystem II

Photosystem II of the photosynthetic apparatus of higher plants is made up of three smaller complexes, viz. the core complex, the light harvesting complex, and the water splitting or oxygen evolving complex.

Core Complex

The core complex is made up of a dozen polypeptides. Genes for eight of these have been localised in the large single copy region of the chloroplast genome. Recently, Nanba and Satoh have made preparations showing P680 reaction centre activity containing three polypeptides, namely, D1, D2 and cyt b559, D1 and D2

Table 7.1. Gene(s) for the Proteins Constituting the Photosynthetic Membranes in Higher Plants

Components of Supra molecular Complexes	*Coding Compartment*	*Gene Designation*	*Clones for Nuclear encoded Genes*
PS II Core Complex			
51 kDa	Chloroplast	psbB	
44 kDa	Chloroplast	psbC	
34 kDa.D2	Chloroplast	psbD	
32 kDa.D1	Chloroplast	psbA	
24 kDa	Chloroplast	psbG	
10 kDa.Phosphoprotein	Chloroplast	psbH	
9 kDa.Cyt b_{559}	Chloroplast	psbE	
7 kDa	?		
6.5 kDa	?		
5.5 kDa	?		
5 kDa	?		
4 kDa.Cyt b_{559}	Chloroplast	psbF	
PS II Antenna Proteins			
Chlorophyll a/b binding apoproteins (24-27 kDa)	Nulceus	cab	Genomic
Water-Splitting Complex			
33 kDa	Nucleus		cDNA
23 kDa	Nucleus		cDNA
22 kDa	?		
16 kDa	Nucleus		cDNA
10 kDa	Nucleus		cDNA
Cytochrome b6lf Complex			
Cytochrome f	Chloroplast	petA	
Cytochrome b6	Chloroplast	petB	
Rieske-FeS Protein	Nucleus	petC	cDNA
Subunit IV	Chloroplast	petD	
PS I Core Complex			
68 kDa Ia	Chloroplast	psaA	
68 kDa,Ib	Chloroplast	psaB	
22 kDa II	Nucleus		cDNA
19 kDa III	Nucleus		cDNA
18 kDa IV	Nucleus		?

16 kDa V	Nucleus		cDNA
12 kDa,VI	Nucleus		cDNA
9 kDa,VII-FeS Protein	Chloroplast	psaC	
PS I Antenna Protein			
Chlorophyll a/b binding apoproteins (22-25 kDa)	Nucleus	Genomic	
Other Components of the Electron Transport Chain			
Plastocyanin	Nucleus	petE	Genomic
Ferredoxin	Nucleus	petF	cDNA
Ferredoxin -NADP$^+$ oxido-reductase	Nucleus		cDNA4
CF1.ATP Synthase			
58 kDa, α Subunit	Chloroplast	atpA	
57 kDa, β Subunit	Chloroplast	atpB	
38 kDa, γ Subunit	Nucleus	atpC	cDNA
25 kDa, δ Subunit	Nucleus	atpD	cDNA
14 kDa, ε Subunit	Chloroplast	atpE	
CF, ATP Synthase			
18 kDa, I Subunit	Chloroplast	atpF	
16 kDa, II Subunit	Nucleus	atpG	cDNA
8 kDa, III Subunit	Chloroplast	atpH	
19 kDa, IV Subunit	Chloroplast	atp I	

proteins show homology and conservation of residues with those of bacterial reaction centre proteins and might be responsible for binding P680, QA, QB, pheophytin and other electron transfer intermediates such as A. D1, the herbicide-binding polypeptide, is encoded by the gene psbA and is one of the first to be localised on chloroplast genome. The gene encodes a polypeptide of 353 amino acids; but the latter is known to get post-transitionally modified possibly by removal of a stretch of 12-16 amino acids at the carboxyterminal end. Sequencing of this gene from atrazine type herbicide-resistant plant species has shown a change from A to G resulting in replacement of the amino acid serine by glycine at position 264.

The gene psbD encodes the polypeptide, D2, of 353 amino acids and, like D1, it is highly hybdrophobic with several membrane-traversing spans. As is well known, a cytochrome,cyt b559, is associated with PSII. This is now known to be made up of two polypeptides of 9 and 4 kDa encoded by genes psbE and psbF. Tha: cytochrome b559 is a product of psbE and psbF genes was predicted after sequencing a fragment containing the gene for psbE which revealed the presence of an ORF for psbF.

Later research proved that this was indeed true. Although the genes for psbE and psbF are highly conserved in spinach, tobacco. *Oenothera*, and wheat, the exact role of cyt b559 remains an enigma. Two genes, psbB and psbC, encode for 508 and473 amino acid long polypeptides, commonly known as the 51 and 44 kDa polypeptides, which bind chlorophyll *a* and are possibly part of the antenna pigment system involved in transferring the light energy to P680 . Both polypeptides show homology and are predicted to have several spans traversing the membrane. The gene psbC at its 5′ end overlaps and 3′ end of another gene, psbD, by 50bp. The genes, psbG and psbH, for two other proteins associated with PSII have also been identified and characterised recently. These two genes encode polypeptides of 248 and 73 amino acid residues, respectively. The product of psbH is a phosphorylated protein.

Light Harvesting Complex

At least four polypeptides of the light-harvesting complex of photosystem II are encoded in the nuclear genome by a family of 3-20 cab genes which can be divided into sub-families on the basis of sequences homology. Furthermore, these genes have also been localised on individual chromosomes. In pea, chromosome two contains a locus for the cab genes whereas in tomato the genes are distributed on at least 5 chromosomes at 5 different loci:cab-1 (chromosome2), cab-2 (chromosome 8), cab-3 (chromosome 3), cab-4 (chromosome 7) and cab-5 (chromosome 12). The locus cab-1 contains four genes, arranged in tandem, whereas locus cab-3 contains three genes and a truncated gene.

The nucleotide sequence of these genes show an open reading frame with a coding capacity of 264-269 amino acids for the precursor proteins. Of these, the first 33-36 amino acids represent the transit peptides which help in intracellular and possibly also intraorganellar targeting of the mature proteins and which get cleaved off by processing protease(s). Most of these gene(s) contain no intron. But, one gene from *Lemna* has been found to contain an 84 bp intron with certain features of a transposable element. Possibly, the intron is not an ancestral feature of the cab genes and has got introduced into the gene later during evolution.

Recently, intron containing cab genes have also been localised in *Petunia* and tomato. They encode slightly divergent, type II, polypeptides which are also incorporated into the light-harvesting complex of PSII.

WATER-SPLITTING COMPLEX

At least five polypeptides have been found of be associated with the water-splitting complex of PSII. Genes for some of these proteins have been localised in the nucleus. Recently, cDNAs encoding the complete precursor for 33, 23, 16 and 10 kDa proteins have been sequenced and found to encode 331, 267, 232 and 140 amino acids which contain aminoterminal transit sequences of 84, 81, 83 and 41 amino acids, respectively. Of the four polypeptides, three (33, 23, 16 kDa) are located inside the thylakoid lumen; hydropathy plots show their hydrophilic nature. Nonetheless, the transit peptides do not show any similarity with those of others at the primary amino acid sequence level. However, they do share among themselves and with the transit peptide of plastocyanin–which is also located towards thylakoid lumen–a unique hydrophobic domain, just before the final processing site, and a region which can form an amphipathic β-sheet and possibly includes the site for primary processing. Secondary structure predictions for the 10 kDa protein suggest one transmembrane segment with the N-terminus in the thylakoid lumen and a 17 amino acid C-terminal segment outside. The transit sequence does not show similarity at the primary sequence level with those of the three extrinsic, luminal proteins of PSII.

Cytochrome b-f Complex

This is the simplest of the supra-molecular complexes associated with photosynthesis and consists of at least four components which are involved in electron flow for PSII to PSI and in cyclic-photophosphorylation. Polypeptides for three of its components, *viz* cyt f, cyt b6 and subunit IV are encoded by the genes pet A, petB and petD respectively, located in the large single copy region. Gene petA, for cytochrome f has been shown to code for a pre-apoprotein of 320 amino acids in several species except for *Oenothera* where the pre-apoprotein is smaller by two amino acids. Comparison of the deduced amino acid sequence with that of the N-terminal sequence of the mature protein has revealed the presence of an N-terminal presequence of 35 amino acids, which gets cleaved off, possibly during transport of the protein across the thylakoid membranes. A topological model of the arrangement of this polypeptide has been proposed on the basis of hydropathy analysis and partial proteolytic cleavage data; it appears that there is only a single span of alpha helix traversing the membrane and the bulk of the amino-terminus polypeptide lies in the intrathylakoid space.

Cytochrome b6 and subunit IV are encoded by two closely placed genes, petB and petD. Recently, Northern blot, reverse transcription, and SI nuclease analysis have shown that both pet B and petD contain introns (of 753 and 742 bp, respectively) and produce polypeptides of 215 and 160 amino acids, the smaller exons code only two and three residues of the amino terminus. Earlier, cytochrome b6 and subunit IV polypeptides have been shown to have homology with the amino and carboxy-terminal ends of mitochondrial cytochrome b, respectively. They have several histidine residues which have been implicated in haem binding. The hydropathy plots show five and three membrane spanning regions for cyt b6 and subunit IV, respectively.

The Rieske-FeS protein has been found to be encoded in the nucleus and a cDNA has been sequenced recently which encodes a complete precursor polypeptide of 247 amino acids, including also a putative N-terminal transit sequence of 68 amino acids. Comparison of the deduced sequence of the mature polypeptide with sequences of the Rieske-FeS proteins of the respiratory electron transport chains of bacterial and mitochondrial origins shows remarkable similarities in secondary structure and the region involved in binding of the Fe_2S_2-cluster. The transit sequence shows some similarity to that of 10kDa polypeptide of the water-splitting system and the delta subunit of ATP synthase complex.

Photosystem I

The core complex of PSI is constituted of at least eight polypeptides. So far, genes for only three of these have been characterised and these are located on the chloroplast genome: four others are located in the nucleus and the location of the remaining one is still unknown. Of the three genes located on chloroplast two psaA and psaB encoding 750 and 734 amino acid polypeptides (and which are substantially homologous) are located in the large single-copy region of the chloroplast genome. These genes possibly arose by duplication. Their products constitute the photosystem I reaction may traverse the membrane. More recently, a third gene psaC, encoding a 9 kDa polypeptide, has been identified in the small single-copy region of the tobacco plastome. The polypeptide possibly functions as the apoprotein for the iron sulphur centres A and B and shows three regions, rich in cysteine residues, with a potential of traversing the membrane.

The light-harvesting complex of photosystem I contains several polypeptides of 22-25 kDa. Recently, one cDNA and one genomic clone (representing however another gene of the family) have been obtained from tomato these genes have been located on chromosome 5 at a distance of about 7 kbp. The gene represented in the genomic clone contains three small introns of 107.87 and 89 nucleotides and the encoded polypeptide has 246 amino acid residues. There is more than 55% divergence in the gene for PSI with members of PSII from tomato.

Genes for other Proteins Associated with Electron Transport

Plastocyanin, a soluble type I copper protein, is a component of the photosynthetic apparatus that transfers electrons from cyt b/f complex to PSI and is located towards the luminal face of the thylakoid membranes. Recently, cDNA and genomic clones for this have been isolated and characterised in spinach and *Silene*. The gene, pet E, encodes a precursor of 168 amino acids with a transit peptide of 69 residues. The mature polypeptide is hydrophilic. The transit peptide shows structural similarity with corresponding regions of those members of the water-splitting complex which are located on the luminal face as discussed earlier. Available data show that the protein derives from a single copy gene which does not contain any intron.

Two other proteins associated with photosynthetic electron transport are ferredoxin-NADP$^+$–oxido-reductase (FNR) and the stromal protein ferredoxin. Genes for these are located in the nucleus and cDNAs, encoding complete precursor proteins, have been sequenced. The ferredoxin cDNA encodes a polypeptide of 146 amino acids comprising 48 amino acids of the transit peptide whereas the cDNA for FNR encodes a polypeptide comprising 55 amino acids of the transit peptide and 314 amino acids of the mature protein.

ATP-Synthase

The ATP-generating supra-molecular complex, ATP synthase, is made up of nine subunits, of which four are part of the membrane-bound sub-complex CFo and five of the extrinsic sub-complex CFI. Subunits, α, β and ε of CFI and I, III, and IV, of

CFo are encoded by the genes atpA (507 codons), atpB (498 codons), atpE (134 codons), atpF (184 codons), atpH (81 codons),and atpI (247 codons), respectively, which are located in two clusters (atp B, E; atpI, H, F, A) about 40 kbp away in the large single-copy region of the plastome (63, 64 and references cited therein). The atpF has been found to be split by a single intron in wheat spinach tobacco and pea. A comparison of the sequence of the actual protein product of atpF with that deduced from the gene sequence reveals that a 17 amino acid peptide from the N-terminal end gets cleaved off before integration. The genes, atpB and atpE, overlap by 4bp in several species though not in pea.

The genes for two subunits γ and δ of CFI and subunit II of CFo are designated as atpC, atpD and atpG, respectively and these have been found to be located in the nucleus. A cDNA encoding the complete precursor (257 amino acids) containing a 70 -residue long transit peptide for the δ subunit, has been sequenced recently. The transit peptide shows some similarity to that of the Rieske-FeS protein, but differs from those of the luminal proteins.

A remarkable feature of the amino acid sequences deduced from these genes is the high degree of conservation not only among different plant species, but also with corresponding genes from bacterial sources.

Nuclear-Plastid Genetic Complementation

The presence of different genomes in a plant cell raises the question : where is the information for biosynthesis of structural and functional chloroplast-specific proteins encoded? What are the contributions of plastid and nuclear genomes to chloroplast biogenesis and to realization of diverse photosynthetic reactions?

It is clear that chloroplast DNA is of utmost importance for plastid self-organization, replication, and genetic continuity. This is well illustrated by Schiff's experiments on the elimination of *Euglena gracilis* chloroplast DNA by streptomycin treatment. This antibiotic specifically blocks chloroplast DNA replication and damages the normal cycle of algal development. As a result of cell multiplication, in subsequent generations the strains formed are devoid of chloroplasts and plastid DNA. Some *Euglena* forms, such as the mutant W_3 *bul*, completely lose their greening capability and are converted from phytoflagellates to zooflagellates. This experiment confirms that plastid DNA is essential as a master molecule for chloroplast formation and development.

Plastid genome mutations cause different kind of hereditary abnormalities in chloroplast structure and function. Phenotypically, plastome mutants are manifested by the appearance of variegated chimeras with sectorial and periclinal localization of green tissues. Unlike virus mosaic, which causes the boundaries between pale and pigmented parts of a leaf to be smeared, the plastome variegation is characterized by a regular distribution of pale and green sectors or by white edging of the full green center of a leaf. Such a picture of variegation is caused by a sorting out of normal and mutated plastids in the process of cell mitosis and tissue differentiation. "Mixed" heteroplastid cells with normal and mutated chloroplasts are found at the boundaries of green and white parts of variegated leaves.

Plastome mutants are extremely interesting for evaluation of the amount of information and for mapping the chloroplast genome. In nature they appear spontaneously with a low frequency of 0.02-0.06%. Plastome mutations can be induced by different physicochemical mutagens.

Considerable progress in the investigation of genetic control of chloroplast biogenesis has been made with the model object *Chlamydomonas reinhardi*. Levine and Goodenough Sager and Kvito *et al* have induced a wide spectrum of mutations in *C. reinhardi* affecting chloroplast properties. Unlike higher plants in which the mutation blocking plastid function is lethal in *C. reinhardi* the cells with inactive chloroplasts can be kept viable by growing the culture in heterotrophic conditions. Genetic analysis of acetate-requiring, antibiotic-resistant *C. reinhardi* mutants led Sager and Rammanis to the discovery of cytoplasmic gene recombination and segregation. They have concluded that in haploid vegetative cells of *C. reinhardi* the plastome is represented by a "diploid" system.

Studying there combination frequency and cytoplasmic gene segregation in repeatedly marked *C. reinhardi* clones, Sager tried to define the linkage group of genophores associated with chloroplasts. This made it possible to start cytoplasmic gene mapping. It turned out that the sequence of linkage groups in cytoplasmic genes represents a ring, which is in agreement with the circular configuration of the chloroplast DNA.

Analyzing the sequence of Mendelian genes affecting chloroplast properties in *C. reihhardi*, Levine and Goodenough suggested that 8 from 16 linkage groups of genes are localized in chloroplast genophores. However, the cytological data of McVittie and Davies do not support this point of view. These authors found 16 chromosomes in the metaphase and related them to 16 linkage groups of nuclear genes. According to Kvitko 16 from 18 linkage groups of genes are localized in the nucleus, 2 in cellular organelles, with one of these two being inherited uniparentally. It is very probable that in meiotic division of *C. reinhardi* cells the transmission function is fulfilled not by all genophores but by single molecules.

It should be noted that the task of mapping chloroplast genophores is very complex. Efforts will be required to elucidate quantitative localization and identification of plastogene properties and functions.

At present attempts are being made for restriction analysis and mapping of chloroplast DNA fragments. Using endonuclease-restrictase, Bedbrook and Bogorad succeeded in splitting a circular DNA molecule of 85 × 10^6 daltons isolated from *Zea mays* chloroplasts. The analysis of DNA fragments by transcription of RNA copies showed that nearly 15% of chloroplast genome sequences were repeated. Two copies of repeats were in an inverted orientation and had 10% of non homologous sequence. The inverted repeats contained genes coding for 23S and 16S chloroplast rRNAs.

Tewari and Wildman noted that pea chloroplast DNA rapidly renatured after heat denaturation, which indicated the presence of a large number of repeated nucleotide sequences.

The high level of homologous sequences in plastid DNA chains points out that the amount of genetic information is limited. According to Smillie, Scott and Bishop the chloroplast genome contains about 2 × 10^7 nucleotide pairs. However, proceeding from the molecular weight of the chloroplast DNA (1-2 × 10^8 dalton), the base content must not exceed 1 × 10^6 pairs. Hence it is possible to calculate the number of genes in a chloroplast genome. Admitting that about 1500 dase pairs from one gene, the chloroplast genome must contain about 1.3 – 1.5 × 10^3 genes. Taking into account the existence of frequently repeating homologous chains, intercistron spacers, and regulatory genes, we can suppose that the chloroplast genome codes for about 150-300 specific proteins of 50 × 10^3 daltons.

In order to determine the nature of chloroplast genetic information, the hybridization of DNA with different types of plastid and cytoplasmic RNA has been successfully used. It was shown that the plastid DNA has a high level of hybridization with rRNA and tRNA extracted from chloroplasts. In the chloroplast genome, 30 -40 cistrons were indentified for each 23S and 16S rRNA. Probably each copy of the plastid DNA contains one gene for the 23S rRNA and one for the 16S rRNA. As for tRNA specific for different amino acids, they are coded for by different genophores. Cistrons of rRNAand tRNA make up about 4% of chloroplast DNA.

The transcription of mRNA onto chloroplast DNA was studied using ^{14}C and ^{3}H uridine. It was shown that the formation of chloroplast mRNA is light induced. In our chloroplast RNA (cRNA) with 9-IIS which upon incubation with ribosomes in vitro makes a "working" polysome. In the presence of the specific inhibitor rifampicin, the light-induced synthesis of all three types of RNA was repressed.

It should be pointed out that not all the components of the PSS of chloroplasts are "products"of their own genes. Schweiger and co-workers transplanted the nucleus from *Acetabularia mediterranea* into an enucleated *A. cliftonii* cell, and showed that hybrid progeny was similar to donor species by the composition of chloroplast ribosomal proteins was demonstrated by Bogored *et al*, in *C. reinhardi* mutants. In higher plants (hybrids of *Nicotiana*), Wildman *et al.* found that 50S subunits of chloroplast ribosomal proteins were not inherited by maternal line. On this basis the authors concluded that the information on the composition of chloroplast ribosomal proteins resides in the nuclear genome.

Chloroplast aminoacyl-tRNA-synthetases are also coded for by the nuclear genome, translated on cytoplasmic ribosomes and then compartmentalized within the chloroplasts. All these data are evidence that the formation of the PSS of chloroplasts is controlled by complementary cooperation of nuclear and plastid genes.

The two genetic systems closely interact in the processes of chloroplast biogenesis, photosystems, and synthesis of Calvin cycle enzymes. Electrophoretic separation of chloroplast membrane polypeptides by SDS acrylamide gel electrophoresis enabled one to identify more than 40 protein bands in thylakoids. A substantial part of these lamellar proteins is coded for by the nuclear genome, as shown in recent experiments by Bingham and Schiff with light-grown *E. gracilis* wild type cells (possessing normal chloroplasts) and dark-grown cells with proplastids and the mutant *W_3 bul* devoid of detectable plastid DNA. In greening cells, five new polypeptide bands were formed in the light. The authors suggested that they were products of plastid DNA genes, Vasconcelos has noted the appearance of nine labeled polypeptides of the thylakoid membranes in isolated *Euglena* chloroplasts. Despite the fact that the number of plastid lamellar proteins is not large, they play an essential role in membrane proliferation, aggregation, and packing of the pigments in thylakoids. It is well known that chlorophyll biosynthesis is controlled mainly by the nuclear genome.

Experimental mutagenesis and induction of the so-called chlorophyll mutation is widely used in investigations of genetic control systems of chloroplast biogenesis. A wide spectrum of mutations which block different stages of synthesis of chlorophyll, yellow pigments, photosynthetic ETC components, Calvin cycle enzymes, and low-molecular weight compounds (amino acids, vitamins,etc) modifying chloroplast morphogenesis, was induced by means of physical and chemical agents (X and γ rays, protons, α particles, laser radiation, ethylmethane sulfonate, nitrosoalkylurea, and others).

Phenotypic manifestation of chlorophyll mutants depends on the mutation nature, whether this is gene or chromosome aberration. It is possible to observe forms from completely bleached up to dark green which differ also in their viability. If the mutation blocks the formation of ETC components or key enzymes of CO_2 assimilation, then the mutants are incapable of autotropic life and they are lethal. Another group of mutations affects only the pigment system of chloroplasts, so their cycle of photosynthetic reaction is not damaged. This group is used for genetic mapping of biosynthesis of chlorophyll and formation of its native forms.

According to Walles' classification nuclear-induced plastome mutant are subdivided into three groups: (a) mutants of one colour–white, yellow, pale green; (b) variegated plants of two or more colours–*alboxantha*, *xantha alba*, *albovirdis*, *virido albina*, *tigrina*, *maculata*, etc; (c) mutants changing the leaf colour during ontogenesis from pale green to a yellow or from white to green, which is connected with photobleaching or chlorophyll stabilization.

Gustafsson was the first to obtain a large collection of barley chlorophyll mutants. Later on, 240 nuclear genes from barley affecting plastid pigment biosynthesis were discribed.

Gottschalk obtained 176 mutation types form pea with genetic blocks in chlorophyll and carotenoid biosynthesis. Using cytogenetic analysis of pea mutants and *chlorina* type, Blixt identified 15 loci controlling the formation of plastid pigments. Wettstein et al.induced 198 recessive lethal chloroplast mutations in barley. Genetic analysis revealed 86 nuclear genes controlling chloroplast development. Five of them (*xantha* - f, g, h, l, u) were classed with structural genes controlling the conversion of uroporphyrinogen to proto-chlorophyllide; 3 genes (*tigrina* 12, "infrared" 5 and 6) were identified as regulatory ones. Among 33 *albina* mutants, allelism was found in only two cases.

Unlike structural genes, two regulatory genes exhibited a similar dominant effect in wild type and mutant alleles in heterozygotes. Using the model plant *A. thaliana*, we have found 60 nonallelic genes controlling chlorophyll biosynthesis and affecting chloroplast ultrastructure and function. As essential problem in studying the biosynthesis of chlorophyll is the elucidation of regulatory mechanisms in the development of its native forms. Accumulation of a definite amount of chlorophyll is to sufficient for normal functioning of photosynthetic membranes. Specific structural organization of chlorophyll into pigment protein complexes (PPC) responsible for primary photochemical processed of light energy conversion in photosynthesis is required. Even at the first hours of greening of etiolated seedlings when the total concentration of chlorophyll is far from being normal, its aggregated forms appear which differ in their spectral properties.

The membrane lipoproteins carrying native chlorophyll forms also play an important role in the molecular organisation of pigment systems. Pigment aggregation and packing in photosynthetic membranes are determined by lamellar proteins, synthesis of which is under the dual nuclear and plastid control. Complementary cooperation of nuclear-plastid genetic systems ensures the development of photosynthetically active membranes, formation of pigment-protein photosystem (PS) assembly, and ETC components. Mutation of both nuclear and plastid genes causes, defects in PS I and PS II. Heber and Gottschalk have described a mutant of *Vicia faba* with a block in PSI.

Chloroplasts of this nuclear mutant are incapable of NADP photo reduction. Bishop induced a mutation in *Scenedesmus obliquus* with a block in the synthesis of cytochrome *f*-552, which caused a suppression of cyclic photophosphorylation. A defect in PS II caused by mutation of one nuclear gene was found by Voskresenskaya,

Drozdova and Gostimskii in the pea lethal mutant. Yakubova *et al.* discovered a block in the formation of ETC components between PS I and PS II in the cotton nuclear mutant. On the other hand, the mutation of chloroplast DNA also causes defects of photosynthetic ETC. Investigating photochemical activity of chloroplasts from *Oenothera* plastome mutants, Fork and Heber and Hallier found that some mutants had defects in PS I and others in PS II. Hagemann *et al* studied chloroplast photochemical reactions in two plastome mutants with damaged photosynthesis. It turned out that in *Antirrhinum majus eniviridis* the mutation of chloroplast DNA caused a defect in PS II.

As a result, they lost ability for the Hill reaction. PS I in this mutant was normal; it did not differ from the wild type by the character of light-induced ESR signals in leaves indicating the normal state of P 700. On the contrary, a damage of PS I and normal functioning of PF II were observed in the mutants 'Mrs. Pollock' of *P. zonale.* This defect correlates with the loss of specific proteins of PPC I. Dissection of genetic systems controlling chloroplast biosynthesis requires a combination of methods of mutational analysis with investigations using specific inhibitors of cell organelle transcription and translation.

Regulation of Expression

In this section, we shall briefly touch upon the regulatory aspects of expression of genes for the photosynthetic membranes. As mentioned earlier, these genes are located both in the chloroplast and nucleus demanding a tight, coordinated, expression which is also evident from the incompatibility of nuclear and plastid genomes in certain inter specific hybrids as studied in detail in the genus *Oenothera.* Although the molecular basis of various malfunctions (e.g. yellowing of hybrids) is still unknown, it is possible that these may arise either from altered signals, and thus breakdown of communication between chloroplasts and nuclei, or the failure to reassemble functional photosynthetic complexes due to improper juxtaposition.

Quite apart from the problem of nuclear-organelle interaction expression of the various genes is also affected by developmental and environmental stimulis particularly light. A proper understanding of various control mechanisms has not developed yet, although some insight has been gained by certain recent investigation as summarised below.

Regulation of Genes of Chloroplast DNA

One extremely important finding is that the cpDNA often codes for a large primary transcript having information for more than one gene which is then processed to smaller units. Such organisation of information is similar to that of operons in bacteria and is reflective of the prokaryotes ancestry of the chloroplast. Structurally also upstream regulatory sequences of chloroplast genes resemble those from prokaryotes in having the–10 and the–35 canonical sequences. At the 3′-end, the precise location of the processing region (i.e. where termination of mRNA may occur)is not clear although several sequences in the non-conding 3′ region show the capacity to form stem and loop structures as in prokaryotes. No polyadenylation of transcripts takes place in chloroplasts. One reason for the very limited understanding of mechanisms of regulation of expression of chloroplast genes is the non availability of chlropolast transformation methods. However, as an alternative, *in vitro* transcription systems have recently been developed for chloroplasts and are being employed to transcribe plastid genes and hopefully rapid progress will be made in the near future.

We have remarked above that light strongly influences plastid development through stimulation of synthesis of certain proteins. This provides a special opportunity to investigate the control mechanism of gene expression. Recent studies have shown, however, that mRNA for most of the chloroplast encoded genes are already present in the dark. Their levels are increased by light, but this increases is rather marginal and much less than of the several proteins which rapidly accumulate in light. By run-on transcription assay, it has been demonstrated that light does not affect relative transcription activities of different plastid genes. Thus, light seems to exercise its effect mediated not so much at the transcriptional but at the post-transcriptional or translational levels.

The best opportunity for a study of events associated with the processing of the primary polycistronic transcripts is provided by the psaA : psbC, psbB: psbH: petB: petD and atpI: atpH: atpF: atpA operons. An understanding of such control is important since of the multiple polypeptides encoded by a single transcript (4 in case of psbB operon), one or more may accumulate even in the dark and others may need light. An intensive investigation into the processing of the transcript of the psbB operon (containing also the psbH, petB and petD genes)has shown a highly complex mechanism. Here, surprisingly, the processing takes place even in the dark, indicating that further controls, including those mediated by light, must operate at the post-transcriptional, translational or post-translational levels. However, we know very little about such controls and unravelling them is a task for the future.

Regulation of Nuclear Genes

As mentioned earlier, genomic clones for only three polypeptides, *viz* chlorophyll *a/b* binding polypeptides of PSI and PS II and plastocyanin, have been characterised so far and they contain the typical eukaryotic expression signals in the 5′-upstream and 3′- untranslated, downstream, regions. Involvement of light in regulation of nuclear-encoded genes became apparent with the studies of Apel and Kloppstech. Subsequently, with the availability of specific gene probes, it was found that both quantity as well as quality of light greatly influence the transcript levels for cab genes, also various members of the gene family are expressed differentially in various organs and cells.

In an effort to identify the elements involved in such regulation, various 5′-upstream regions of the cab gene have been tested by transforming a heterologous system with engineered genes in which either deletions have been made in the 5′-upstream region or various parts of this region have been fused to a marker gene containing its own constitutive type of promoter. Simpson *et. al.* reported that in pea a 400 bp fragment from the cab gene is responsible for light-regulated and organspecific expression as is evident from experiments involving fusion between this fragment and a modified bacterial kanamycin resistance gene–both engineered into a Ti plasmid–and subsequently employed for transformation of tobacco. Out of this, a 247 bp (–100 to –347 bp upstream) stretch acts not only as a light-regulated enhancer element but also as tissue specific silencer.

Lamppa *et. al.* have reported that a wheat cab gene maintains its light-regulated characteristics even after transfer to the more distantly related tobacco and *Petunia* indicating that the controlling elements are ubiquitous in flowering plants. A 5′-upstream region of this gene from + 31 to –1816 bp has been shown to confer light-regulated and organ-specific expression also on a bacterial chloramphenicol resistance gene. Expression of this gene is in fact, affected reversibly by red/far-red light implying a role of phytochrome in the regulation of its expression. Further experiments have shown that only a 268 bp fragments from –89 to –357 bp in the 5′-upstream region, which acts as an enhancer, is sufficient to confer the typical regulatory characteristics. The work mentioned above has set the stage for more detailed investigations on the mechanism of action of the enhancer elements. Such sequences from the rbcS gene have already been used to identify, by gelretardation and foot-printing methods, possible trans-acting factor(s) responsible for light -regulation as also organ-specific expression of the genes. It is hoped that similar work with the cab gene in the near future will further elucidate the pathway and mechanism of transfer of a stimulus to a gene.

As in the case of cab and rbcS genes, ,mRNA levels of many other nuclear-encoded genes are greatly increased by illumination. However, the quantity by which mRNA levels increase is often greater than that of the increase in level of proteins for several components of the photosynthetic apparatus–e.g., mRNAs of the delta subunit of ATP synthase, the Rieske iron sulphur protein of the cytochrome b/f complex, and the 33, 23, and 16 kDa polypeptides of the water- splitting system. This indicates again that there must exist novel post transcriptional control mechanisms for further regulation. Such controls may lie at the translational level or even at the level of transport of precursor proteins into the chloroplast (the latter may in turn depend on the availability of receptors) for final localisation in the supra-molecular complexes. All these putative mechanisms of control remain to be elucidated.

Physiological and Genetic Basis of Improving Photosynthetic Productivity

Analysis of advances in modern agriculture when technical progress of husbandry has reached a high level shows that an increase in the productivity of the main crops has a tendency to stabilization. Further elaboration of agricultural technology and considerable increase in mineral nutrition dosage does not cause corresponding yield enhancement. Does it mean that all the productivity potential of modern plant growth has been exhausted and maximum yield for some crops achieved?

Basic physiological and genetic investigations of plant productivity show that the yield potential is rather high and not yet fully exploited. A rich gene pool in crop plants has not been fully utilized for the purposeful and efficient enhancement of those processes which at present limit crop productivity. Among the yield-limiting factors, we should mention first of all photosynthesis, the efficiency of which is extremely low. Efficiency of solar energy utilization for many crops does not exceed 1% . Moreover, cultivated varieties of such plants as wheat, cotton, and others have lower rates of photosynthesis that their initial wild types. However, high productivity of crops has been provided by leaf area increase, changes in the ratio of the biomass of reproductive organs to that of vegetative ones, and by other morphophysiological properties. It is quite evident that plant breeding research has not been aimed at improvement of the photosynthetic apparatus.

Mass selective screening has been conducted according to the morphological properties and traits of agronomic value, with much attention being given to increasing the size of sink and storage organs- the number and size of ears, bolls, tubers, etc. Enhancement of the plant sink and storage capacity resulted in an increase of the agronomic yield as long as the photosynthetic potential permitted. At present an imbalance can be observed between the desirable a number and the size of economically valuable plant organs and the possibility of providing them with the products of photosynthesis in the newly obtained promising varieties and hybrids. This imbalance between the accumulation and partition of assimilates leads to puniness and immaturity of seeds, falling of ovaries and bolls, decrease of sugar content, and other disorders. Besides, and excessive increase in the sink power can provoke, by the negative feedback mechanism, a decrease in the activity of the photosynthetic apparatus due to the with drawal of nitrogen from leaves into other organs thereby causing their premature senescence.

From the above we can conclude that low photosynthetic efficiency became a "bottleneck" for productivity enhancement. Development of methods for photosynthesis regulation and the increased efficiency in utilization of solar energy up to 3-5% are the most important means of obtaining high agricultural yield. The world's experience in obtaining highly productive varieties and high yield on the basis of modern agricultural technology shows that the possibilities of crop yield enhancement by the best utilization of water, fertilizers, light energy, and other factors are related to the optimization of canopy structure and improvement of the photosynthetic apparatus on physiological and genetic bases. It is calculated that an increase of photosynthetic utilization of photosynthetically give radiation (PAR) up to 3-5% will permit the following crop yields : maize–15 tons of grain, winter wheat–10 tons of grain, seed cotton–8 tons per hectare. The relationship between photosynthesis and crop yield is very complex. In the literature there are contradictory data on the correlation between photosynthetic rate and crop yield.

The absence of a direct relation between them is probably explained by the dependence of crop yield on the net assimilation value which in turn is determined not only by photosynthetic rate but also by leaf surface, size, duration of vegetable period, canopy structure, dark and light respiration, translocation, and partitioning of assimilates. An integral scheme of phenotypic expression of nuclear and cytoplasmic gene action in formation of the factors of photosynthetic productivity and crop yield. It is seen from this scheme that crop yield is the result of complex diverse processes and reactions occuring at ontogenesis under the influence of external conditions. Thus, three ways to increase the productivity follow from these conceptions: 1 optimization of the canopy structural as an optical system; 2. improvement of efficiency of the photosynthesis process itself; and 3. rational partition and usage of assimilates for formation to the agronomic yield (harvest index).

The first way, based on the morphophysiological and agroecological data, is widely used in agronomy. The theory of this problem and the principle of canopy structure optimization with favourable leaf area index (LAI) allowing improvement of radiation and aerodynamic conditions for different crops were developed in detail by Nichiporovich and other investigators.

The second and the third ways to increase photosynthetic productivity concern the genetic balance of plant source and sink abilities. According to Evans, it is important to know when source or sink limit the crop yield. In cereals, for example, grain filling is largely dependent on photosynthesis and environmental conditions after flowering, whereas the capacity for storage is determined by conditions before flowering. Nalborczyk and Gej showed that in different varieties of spring wheat the photosynthetic productivity during milk and wax maturity played an important role in obtaining high economic yield. A remarkable pecularity of the intensive varieties was a coincidence of net assimilation rate (NAR) maximum with that of LAI.

The interrelation between source and sink has a pronounced feedback effect. For example, removal of the wheat ear or the cotton boll causes a considerable depression of photosynthesis. On the contrary, increasing sink capacity can enhance to a certain extent the rate of photosynthesis. In experiments by Thorne and Evans grafting of spinach to sugar beet led to an increase in assimilating ability of the graft.

Analyzing the numerous data on the correlation between source and sink capacity, it should be noted that the one-sided approach to these processes does not give positive results. For example, CO_2 feeding usually results in considerable alteration of chloroplast ultrastructure due to an excessive storage of starch. At the same time, a great attractive power can lead to chloroplast "wear" and pressure leaf senescence.

Section of plants on a physiological and genetic basis will make it possible to get varieties and hybrids with a balanced ratio between source and sink that will provide the maximum expression of yield potential for each plant.

Analysis and Mobilization of the Germplasm

Evaluation of the diversity of photosynthetic functions and identification of the parameters of photosynthetic efficiently allow practicing breeders to manipulate the genotype for significant improvement of plant productivity. Ecological and geographical populations of plant species have a large-genetic variability of chloroplast photochemical and enzymic activity which provides their different photosynthetic ability.

The genotypic determination of chloroplast structure and function is clearly manifested in photosynthetic pathways of C_3 and C_4 plants. The plants with Kranz anatomy of the leaf (maize,sorghum,sugar cane,etc) are known to possess an additional pathway of CO_2 fixation with the participation of phosphoenol-pyruvate (PEP) and the formation of C_4 dicarboxylic acids. Hatch and slack earlier proposed that the C_4 pathways are an autocatalytic cycle of CO_2 assimilation through transcarboxylation of C_4 acids. Later it became obvious that the C_4 pathway is an auxiliary channel of primary CO_2 fixation on the Calvin cycle. In all higher plants, irrespective of the type of primary fixation, CO_2 reduction to carbohydrates occurs in the pentosephosphate cycle. However, C_4 plants have the advantage of high resistance to extreme temperature and more efficient use of water, thus adapting better to arid climates.

Besides, C_4 plants are characterized by a CO_2 compensation point close to zero, the photosynthetic rate being independent of the concentration of oxygen since they practically do not have photorespiration. All this, together with high light requirement, provides a very high photosynthetic capacity which is about twice as great as that of C_3 plants. The highest rate of photosynthesis (85 mg CO_2 dm^{-2} hr^{-1}) is observed in maize in natural conditions. Photosynthesis in C_4 plants in natural conditions is not light saturated.

The most important feature of C_4 plants is the presence in their leaves of two types of chloroplasts which differ by their ultrastructure and functions. In palisade mesophyll cells the chloroplasts have a pronounced granal strucure,while in parenchyma bundle-sheath cells they are usually paucigranal. These chloroplasts differ greatly in their photochemical properties : PS II is active in mesophyll cells while PS I is active in bundle-sheath chloroplasts. The key enzyme of the Calvin cycle, RuDP carboxylase, is absent in mesophyll cells. According to Hatch, the primary CO_2 fixation in C_4 plants oxaloacetic acid (OAA). The process proceeds in the cytoplasm since PEP carboxylation which leads to the formation of oxaloaecetic acid (OAA). The process proceeds in the cytoplasm since PEP-carboxylase is a purely cytoplasmic enzyme. In chloroplasts OAA is transformed into malate and aspartate. Further, depending on species, malate or asparate are transported to the bundle-sheath cells where due to decarboxylation of the acids CO_2 is released and refixed in the Calvin cycle.

Cooperation of two types of plastids in C_4 plants leads to a concentration of CO_2 in the bundle-sheath cells where it decreases to a certain extent the limitation of photosynthesis by the diffusive CO_2 resistance and thus aids in increasing the efficiency of the process. On the other hand, all this prevents the CO_2 burst due to photorespiration. Probably the primary function of the C_4 pathway is the reassimilation of CO_2 as a result of this wasteful process.

How can the dimorphism of chloropalsts in C_4 plants be explained from the genetic point of view? Is this a result of sorting out the geneticallly nonhomogeneous plastids originated from the parental pair, or are they formed under the control of the nuclear genome during cell differentiation? We believe that the second assumption is more correct. For example, maize belongs to the group with the maternal type of plastid inheritance therefore it has homogenous initial plastid material. In the process of differentiation of the cells and tissues under nuclear control, mesophyll chloroplasts are specialized for the C_4 pathway. This is probably achieved by a repression of nuclear genes for the small subunit RuDP carboxylase which as mentioned above, plays a regulatory role in transcription of the chloroplast mRNA of the large subunit. It is quite possible that genes coding for the formation of ETC components of PS I (ferredoxin, NADP reductase, cytochrome b-563) are localized, namely on the small subunit chromosome. As a result of this gene repression, the activity in PS I of mesophyll chloroplasts is three to four times lower than that of bundle-sheath cells. Thus the nucleus controls the division of labour between the chloroplasts of mesophyll and bundle sheath cells. However, this hypothesis requires further experimental confirmation.

The C_4 pathway is observed not only in tropical grasses, but it is also widespread among dicotyledons. Within one genus, C_3 and C_4 species and intermediate forms are encountered. It is possible by hybridization to inpart ot C_3 plants the C_4 pathway? Bjorkman *et al.* successfully carried out the crossing and obtained a fertile hybrid between C_3 and C_4 *Atriplex* spp., but the efficiency of photosynthesis was lower than that of either parent in all F_1 and F_2 hybrids. These experiments are currently being continued by Nobs but they have not yet led to desirable results.

The C_4 pathway is connected with the leaf Kranz anatomy, and therefore it is impossible to impart properties of cooperative photosynthesis to C_3 plants. However, we should not exclude the possibility of enhancement of the auxiliary CO_2 fixation through PEP carboxylase. As has already been mentioned, this enzyme operates in the cytoplasm and its formation is controlled by the nucleus. It has a higher affinity for HCO_3^- than for CO_2; therefore it can bind the hydrated CO_2 more efficiently and transfer it to the chloroplsts, thus decreasing the limitation of photosynthesis by carboxylation resistance. The activity of PEP carboxylase in C_3 plants is 60 times lower than that in C_4 plants.

Since the C_4 pathway originated in the course of evolution as a result of plant adaptation to arid and hot climates, it might be assumed that the forms with high activity of PEP carboxylase can be obtained among C_3 plants mutants by genetic screening in rigid condition. An essential part of this assumption is that the cytoplasmic system of carboxylation and CO_2 transport to the chloroplast RuDP of C_3 plants should be provided with the mechanism of acceptor regeneration, i.e., pyruvate conversion into PEP by pyruvate kinase.

Experimental mutagenesis is a powerful means fo reconstructing the genome and breeding new promising forms, which able shifts in the structural and biochemical organization of chloroplasts. As a result, the overwhemling majority of viable mutants have a reduced photosynthetic capacity. There are still a few mutants (1-2%) belonging to the "plus" class as measured by the increased photosynthetic efficiency. Photosynthetically positive mutants were obtained in *Chlorella*, pea and cotton. In research at our institute with γ radiation, 60 cotton mutants were obtained among which there was one promising mutant-*Duplex* with a compact type of branching, high frequency of double sympodial bolls, fast ripening, and high agronomic yield. Its harvest index amounts to 50%, which points to rational partition and utilization of assimilates. The economic yield of this mutant is at the rate of 1 ton per hectare of seed cotton higher than that of the initial variety 108-F.

Physiological and biochemical studies have shown that the mutant *Duplex* is characterized by a higher photochemical and enzymic chloroplast activity during the reproductive period. In this mutant the endogenous delay of leaf senescence is observed, whereas in the initial variety, as well as in many other crops, a decrease in the specific activity of key Calvin cycle enzymes occurs on the 20th - 30th day after the leaf expands, when there are no visible alterations of chloroplast ultrastructure. Physiologic-genetic prolongation of the functionally active state of chloroplasts and prevention of premature leaf senescence are important factors in high photosynthetic productivity.

In the last few years the attention of many researchers has been attracted to the problem of photorespiration. According to the data obtained by Zelitch in C_3 plants (sunflower, soybean, sugar beet, etc), photosynthesis product losses upon photorespiration can reach 50% of the net assimilation. Photorespiration is considered to be a waste of organic compounds which could be used in the productive process. This is evidenced by the data on photosynthetic efficiency of C_4 plants in which photorespiration is practically absent. This circumstance can serve as one of the arguments proving that photorespiration as a process of CO_2 release in the light has no great physiological importance.

The main source of CO_2 release in the process of photorespiration is decarboxylation of intermediates of glycolate metabolism. The glycolate pathway is closely connected with the

Calvin cycle, particularly with oxygenation of RuDP resulting in the formation of the phosphoglycolic acid. It is generally accepted that the key enzyme of the Calvin cycle, RuDP carboxylase, plays a bifunctional role catalyzing carboxylation and oxygenation reactions. Oxygen and CO_2 complete for the reaction substrate RuDP.

In the enzyme oxygenase reaction, the sulphydryl groups take part; their total amount is 96, but the number of easily modified, "accessible" ones is 12. It has been shown recently that the photorespiration inhibitor glycidate irreversibly inactivates oxygenation, the carboxylase activity of the enzyme not being affected. Apparently oxygenation-specific inhibition is connected with blocking of enzymes SH groups located in its active centre.

In its native state the conformation of this oligomeric protein complex and its functional activity depend on conformation and dimensions of large and small subunits. Analyzing the peptide composition of the enzyme in 60 tobacco species and 10 plants belonging to different phylogenetic groups–from blue-green algae to ginkgo–Kung has found that in all species studied the large subunit contains only three polypeptides, while the number of polypeptides in the small subunit varies from 1 to 4.

The latter may serve as a genetic marker in the investigations of nuclear-cytoplasmic interaction. It may be assumed that depending on the polypeptide composition being coded for by the small subunit of the nucleus, the carboxylation or oxygenation activity of the enzyme may vary within plant species. Indeed, varieties with low photorespiration have been found within populations of C_3 species. By means of genetic screening, Zelitch has distinguished tobacco population from the variety 'Havana Seed' in which 25% of the progeny had low photorespiration.

It is essential to determine what methods can be used for mass screening for the desirable forms with low photo respiration. Growing the plants in a hermetic chamber at low CO_2 concentration prior to screening them for viability is one of the simplest methods of mass genetic screening for photorespiration. However, this method has not been successful since plants in the closed chamber were not viable. In this relation the method of photorespiration estimation by CO_2 burst measurement after illumination appears to be promising as do investigations of dependence of NAR on O_2 concentration. In conclusion it must be noted that estimation of photosynthetic efficiency, genetic screeing of "plus" mutants, and the mobilization of species germplasm for the improvements of plant productivity should be conducted, taking into account other physiological properties such as immunity and stress resistance. In the near future, the possibility of genome mapping and genotype manipulation by means of somatic hybridization and gene engineering will allow practicing breeders to create new, synthetic plant varieties with a desirable combination of the valuable traits, high photosynthetic efficiency and maximal expression of economic yield.

8

ALGAL PHOTOSYNTHESIS

Estimates of the rates of photosynthetic energy conversion on regional and global scales are required for predicting the energy available to food chains and for determining the role of algae in geochemical cycles. As all organisms are ultimately dependent on energy input *via* photosynthesis, and limitations to global carbon dioxide assimilation may constrain the potential productivity of higher trophic levels. Riley (1944) provided the first estimate of global oceanic primary productivity of 860 ± 560 mg $C.m^{-2}.d^{-1}$. The large standard deviation associated with this estimate arose from uncertainty in extrapolating a small number of observations from a limited range of open and coastal ocean regions to the entire ocean. Despite an additional 45 years of research involving the widespread incorporation of ^{14}Cassimiation measurements into biological ocean graphic sampling programs, there is continuing uncertainty about the magnitude of oceanic primary productivity.

Shipboard observers cannot adequately sample the ocean at all spatial and temporal scales necessary to adequately resolve variations in phytoplankton biomass and productivity. In contrast, satellite remote sensing offers the potential to provide a detailed assessment of phyhtoplankton spatial and temporal variations. The remotely sensed signal is used to infer chloraphyll *a* concentration from measurements of light reflected from near surface waters. At present, remote sensors do not provide a direct measure of primary productivity; this must be inferred from the relationship between primary productivity and chlorophyll *a* concentration. It has long been a challenge to estimate phytoplankton productivity from chlorophyll *a* concentration, because pigment concentration can be sampled much more rapidly than photosynthesis. Introduction of *in vivo* fluorometry to continuously monitor chlorophyll concentration together with radiometric sensors for estimating chlorophyll *a* from aircraft or satellites and moored fluorometers to continuously measure cholophyll *a* at fixed positions in continental shelf waters have greatly increased the spatial and temporal resolution with which phytoplankton abundance can be determined.

Unfortunately, there has not been a corresponding increase in the ability to measure phytoplankton photosynthesis, even though deployment of high-resolution O_2 sensors, pump and probe fluorometers and sensors to measure passive chlorophyll fluorescence may allow increases in the temporal and spatial coverage of phytoplankton primary production. In lieu of more widespread application of these techniques, investigators have attempted to estimate primary production from chlorophyll *a* concentration and a range of environmental variables, using empirical and analytical algorithms. This chapter considers (1) the methods for remote sensing of the chlorophyll *a* concentration in surface waters and (2) the ways in which chlorophyll *a* distributions can be used to estimate primary production.

Remote sensing of phytoplankton pigment concentration can be achieved by (1) passive measurements of water colour (2) passive measurement of solar-induced chlorophyll fluorescence and (3) active measurement of laser induced fluorescence. Water colour is quantified by measuring the radiance upwelling from the surface waters using shipboards, aircraft, or satellilte radiometers. Satellite instruments currently employ radiometers with broad spectral bands (60-100 nm) in the visible region, limiting measurements to a single pigment (chlorophyll *a*) in clear waters, with a low diversity of optically active components. However, in clear ocean waters, the ocean colour algorithms for remote sensing of chlorophyll *a* yield relatively good precision over a wide range of chlorophyll *a* concentrations without the need for extensive, independent ground truth observations.

Much greater spectral resolution is possible with ship and aircraft borne radiometers with the potential for evaluating the concentrations of several major pigments even in turbid coastal waters with a large diversity of optically active components. However, accurate remote sensing of the phytoplankton concentration in a complex mixture first requires more detailed knowledge of the optical properties of phytoplankton as well as measurements of the other components of these assemblages, such as particulate material and coloured substances. This information becomes more directly applicable to satellite remote sensing during the 1990s with the development and development of satellites that will carry moderate- and high-resolution spectrometers.

Passive remote sensing of *in vivo* chlorophyll *a* fluorescence relies on high spectral resolution measurements of upwelling radiance at wavelengths around 680 nm. The technique is based on the same principals as *in vivo* fluorometry but differs from the *in vivo* technique in two important respects. First, passive remote sensing of chlorophyll a fluorescence relies on a natural, variable light source (the sun) instead of an artificial source with a known spectral quality and irradiance. Second, remote sensing of chlorophyll a fluorescence samples the upwellling irradiance that leaves the water surface rather than a known small volume. The remotely sensed fluorenscence emission is, however, superimposed on a backscatter signal of unknown magnitude which can complicate the interpretation of the fluorescence signal. In addition, the approximately order- of-magnitude variability of *in situ* fluorescence yield.

Active remote sensing of surface phytoplankton abundance relies on the use of lasers pulsed at short intervals to excite *in vivo* chlorophyll a fluorescence and a photomultiplier to measure the induced red (680 nm) fluorescence. The primary fluoroscensor in use today is the Airborne Oceanographic Light Detection and Ranging Detector (AOL) developed at the NASA Goddard Wallops Island Flight Station. In this system, the excitation wavelength in 532 nm and the return signal to an airborne spectrometer includes contributions from back-scattering at the excitation wavelength, Raman scattering by water, and fluorescence by chlorophyll *a* and phycoerythrin. Laser systems for measuring pigment concentrations are limited to low-flying aircraft largely because of the power requirements for operating the laser, atmospheric attenuation of the signal, and bulky instrumentation.

The AOL allows simultaneous measurement of active laser-induced chlorophyll fluorescence, passive solar-induced chlorophyll fluorescence, and passive ocean colour and can be used to obtain simultaneous estimates of chlorophyll concentrations from passive and active techniques. Good correlations between laser-induced chlorophyll *a* fluorescence and ground-truthed chlorophyll *a* concentrations suggest that variations in fluorescence yield are not significant during the approximately 1-2-hr flight time typically employed in experiments using the AOL.

Estimating Surface Water Chlorophyll a

Most phytoplankton abundance measurements to date have relied on ocean colour measurements using the now defunct Coastal Zone Colour Scanner (CZCS) on the Nimbus 7 satellite. Estimating chlorophyll *a* from ocean colour measurements rests on the observation that dissolved or suspended materials change the reflectance of water through the absorption and scattering of light. Radioactive transfer theory serves as the basis for examining the relations among reflectance, absorption, and scattering. A model developed by Gordon *et.al* (1975) leads to the following equation relating subsurface reflectance to the absorption and back-scattering coefficients:

$$R_s(\lambda) = 0.33b_b(\lambda)/[a(\lambda) + b_b(\lambda)] \quad ...(1)$$

Where $R_s(\lambda)$ is the subsurfaces reflectance (dimensionless), at (λ) is the absorption coefficient (m^{-1}) and $b_b(\lambda)$ is the backscatter coefficient (m^{-1}). The absorption and back scatter coefficients with units of inverse length (*i.e.,* m^{-1}) can in turn be expressed in terms of the concentrations and specific absorption or scattering coefficients of optically active material. It is commonly assumed for open ocean waters that most of the backscattering is due to water itself and most of the variable absorption is due to phytoplankton pigments together with substances that covary with phytoplankton abundance, leading to the approximation.

$$R_s(\lambda) = 0.33b_{bw}(\lambda)/[a^*(\lambda)\ C + a_w(\lambda)+b_{bw}(\lambda)] \quad ...(2)$$

where $a^*(\lambda)$ is the typical chlorophyll a-specific absorption coefficient for phytoplankton ($m^2.g^{-1}$ chla) and C is the chlorophyll *a* concentration ($g.m^{-3}$). More complicated expressions, including contributions of dissolved and suspended inorganic matter or dissolved organic substances, such as humic and fulvic acids, are required for near-shore waters and lakes. Changes in the concentration of chlorophyll a will markedly affect reflectance in those spectral regions in which phytoplankton absorb light most strongly (*i.e.,* in the blue and red).

Algorithms for determining chlorophyll a concentration from CZCS images depend on an empirical relationship between the ratio of blue to yellow light upwelled through the ocean surface and chlorophyll a concentration. This relationship depends on the observation that the "typical" phytoplankton absorption spectrum has peaks centered in the blue ($\lambda = 440$ nm) and red ($\lambda = 670$ nm) regions but does not absorb strongly in the yellow ($\lambda = 550$ nm). The algorithm has the form.

$$\log[C] = a + b \log [R(443)/R(550)] \qquad ...(3)$$

where $R(\lambda)$ is the reflectance centered at wavelength λ and *a* and *b* are fitted coefficients obtained from a comparison of surface reflectance derived from satellite measurements corrected for atmospheric effects and sea surface chlorophyll *a* concentrations determined from ship observations.

The approach (Eq. 3) is quasiempiric, resting on the observation that high absorption of blue light by phytoplankton reduces the reflectance in this spectral region relative to wavelengths in the yellow where pigment absorption is minimal. The ratio of reflectance in the green ($\lambda = 520$ nm) to yellow ($\lambda = 550$ nm) spectral bands of the CZCS have also been used to replace R(443)/R(550) in Eq. 3 with equally good precision. Estimates of remotely sensed chlorophyll a are within ±40% of surface observations over a range of approximately 0.1-50 μg chla.dm^{-3}. Limits on the precision are due to errors in making atmospheric corrections, variability in phytoplankton optical properties among diverse phytoplankton assemblages, and variability in the optical properties of seawater due to the nonuniform distribution of dissolved organic matter and suspended sediments.

The reflectance ratio algorithms were developed for use in optically clear (Case 1) ocean waters in which chlorophyll a and substances that covary with chlorophyll *a* are the major sources of optical variability. Use of algorithms based on blue to green reflectance ratios are unlikely to resolve phytoplankton abundance in estuarine waters because of light absorption in the blue and green spectral regions by iron oxides and humic-acid-like material, which can be important is estuarine and fresh-waters.

To overcome this problem, the ratio of reflectances in the red to near infrared, where absorption by dissolved materials is minimal, can be employed. The principal is the same as that employed by Gordon et.al. (1983) for clear open ocean water measurements. High phytoplankton absorption in the red reduces reflectance at this wavelength relative to a region of low pigment absorption in the near infrared. This approach has been used by Stumpf and Tyler (1988) to measure chlorophyll *a* concentrations in excess of 5 μg.chla.dm^{-3} in turbid estuarine waters with a precision of ±60% or 5 μg.chla.dm^{-3}, whichever is the greater.

Although satellite remote sensing provides a means of greatly increasing spatial coverage, the technique is subject to a number of limitations including (1) restricted temporal coverage due to cloud cover, (2) inaccuracies in processing algorithms for making atmospheric corrections, particularly at large zenith angles Pan *et al.* (1981), (3) poor vertical resolution that is limited to the upper 20% of the euphotic zone which includes a small and variable (1% to 8%) portion of chlorophyll *a* in ocean waters.

In addition, the depth interval that is included in a remotely sensed measurement will vary with wavelength (λ), being inversely proportional to the attenuation coefficient K(λ). For example, as the chlorophyll concentration rises from 0.01 to 10 mg.m^{-3}, the depth zone that accounts for 90% of the upwelling irradiance decreases from 58 to 2 m at 443 nm and decreases from 15 to 4 m at 550 nm, while remaining virtually constant at 2 m for 670 nm light. Thus, passive and active remote sensing of chlorophyll a fluorescence is limited to the very shallow depths near the surface, whereas observations of ocean colour depend on light that penetrates deeper into the water column. Other variables, in addition to chlorophyll *a* concentration, that are important in determining photosynthesis rates, and that can be measured by aircraft and/or satellite sensors include temperature and downwelling irradiance at the earth's surface and the depth of the surface ocean mixed layer. These measurements will not be discussed further, but the importance of including these variables in algorithms for the remote sensing of phytoplankton photosynthesis will become apparent in the remainder of this chapter.

Empirical Relationships Between Primary Production and Chlorophyll a

As a first approximation, primary productivity is expected to be proportional to plant biomass, all other factors being equal. Both limnologists and oceanographers have observed linear relations between primary production and chlorophyll *a* concentration for data expressed on either a unit volume or unit area basis, with biomass accounting for 50% to 75% of the variance in primary production. For remote sensing applications, it is desirable to relate integral primary production (with units of $gC.m^{-2}.d^{-1}$ and designated π) to near surface chlorophyll *a* concentrations (with units of mg. chla. m^{-1}, and designated C_s). Despite the linear relation that is sometimes observed between π and C_s, theoretical analyses suggest that the relationship between integral primary production and phytoplankton abundance should be nonlinear largely because of the influence of other light-absorbing substances. In fact, a logarithmic relation is often observed between areal primary production and surface phytoplankton abundance:

$$\log_e(\pi) = \log_e(a) + b \log_e(C_s) \quad ...(4)$$

or

$$\pi = aC_s^{\,b} \quad ...(5)$$

where *a* and *b* are empirically derived coefficients. Observations for various ocean regions are characterized by *b*=0.4 - 0.6. In contrast, Brylinsky (1980) found that annual phytoplankton production scaled as mean chlorophyll *a* concentration raised to the power 1.36 in lakes. Whether this difference is due to the wide variation in the temporal scales over which data were averaged (one day for the oceanic data versus a growing season for the lake data), or because of a bias introduced by other uncontrolled variables, or to a fundamental difference in the relation between π and C_s in freshwater and marine systems remains to be evaluated.

Variations in primary production that cannot be accounted for by differences in surface biomass are presumably due to differences in other factors such as temperature, irradiance, nutrient availability, optical properties of the waterbody, vertical distribution and species composition of the phytoplankton. Fee (1979) found that annual values of π/C_s for lakes in North-Western Ontario varied with the dominant phytoplankton taxta. This ratio of π/C_s was greatest in lakes dominated by members of the Chlorophyceae, intermediate in lakes dominated by members of the Chrysophyceae, and lowest in lakes dominated by cyanobacteria.

Two environmental factors that are amenable to remote sensing and would be expected to influence the relationship between primary production and chlorophyll *a* concentration are irradiance and temperature. To examine the effects of these and other variables on the rate of phytoplankton photosynthesis, it is desirable to define an areal productivity index (π_{chl}), as the ratio of integrated euphotic zone primary production (π) to integrated euphotic zone chlorophyll *a* content (C_i):

$$\pi^{chl} = \pi/C_i \quad ...(6)$$

where

$$\pi = \int_0^D P(z)\,dz \text{ and } C_i = \int_0^D C(z)\,dz_s$$

D is the depth of the euphotic zone, P(z) is the primary production at depth z, and C(z) is the chlorophyll *a* concentration at depth z. A linear relation between the real productivity index (π^{chl}) and incident irradiance (E_0) has been observed by several investigators. Platt (1986) suggested that the slope of this relation, designated (ψ), was related to the chlorophyll a-specific absorption coefficient a*, the realized photon yield of photosynthesis in situ (η), the euphotic zone depth (D), and the vertical attenuation coefficient (K) for scalar irradiance as follows:

$$\pi = \pi^{chl}\, C_i = C_i\, \psi E_0 = C_i a^\bullet\, \eta E_0 DK \qquad ...(7)$$

From a review of published data, Platt (1986) and Platt et al (1988) concluded that ψ varied approximately by a factor of two and, based on theoretical grounds, argued for the relative stability of ψ. In contrast to Platt's (1986) conclusions, Campbell and O'Reilly (1988) found that ψ varied from 0.1 to 10, with a mean value approximately three times higher than obtained from Platt's (1986) summary. Observations of Yoder *et al.*(1985) for the Southeastern United States continental shelfalso yielded a value for ψ, which was about three times greater than the typical values given by Platt (1986) suggesting some regional variations.

Variability in ψ may arise from a number of sources, including (1) variations in a,* (2) variations in η due to variations in the vertical distribution of phytoplankton abundance and in the relationship between phytoplankton biomass and vertical light attenuation, and (3) variations in η that arise from variations in the parameters of the PI curve P_m^{chl} and α^{chl}. It is necessary to consider the full equation for determining π from the depth-dependence of photosynthesis to examine the sources of variability in ψ. Recent theoretical treatments of this problem which explicitly consider the angular and spectral distribution of phytoplankton abundance, are beyond the scope of this review.

One variable that is particularly important in determining ψ is the attenuation coefficient (K). Noting that the euphotic zone depth (D) and attenuation coefficient (K) are related by the equality DK = $\log_e$ (0.01) = 4.61 (*i.e.*, K = D/4.61) allows Eq. 7 to be rewritten as

$$\psi = a^8\eta\ 4.61/K^2 \qquad ...(8)$$

Because of the K^2 dependence of ψ, explicit consideration of the vertical attenuation coefficient may improve models of integral primary production (π). Explicit consideration of the attenuation coefficient may be particularly important in estuaries where K can vary greatly because of changes in phytoplankton abundance, dissolved organic matter, and suspended sediments.

In fact, Cole and Cloern (1987) showed that π scaled with $C_c E_d/K$ in five United States estuaries. A linear dependence of π on $C_s sE_0/K$ has also been demonstrated in Delaware Bay and Narragansell Bay. Significantly Pennock and Sharp (1986) found that the slope was three times greater in the summer than during the nonsummer months. They attributed this difference to a change in species composition of the phytoplankton from predominantly microflagellates in summer to diatoms for the rest of the year.

Table 8.1. Dependence of integral primary productivity (p) on chlorophyll concentration (C*s*), incident irradiance (E*o*), and vertical attentuation coefficient (K).

Location	*Slope*	*Correlation Coefficient*
San Francisco Bay		
North Bay (1980)	0.15 ± 0.026	0.72
South Bay (1980)	0.19 ± 0.026	0.88
South Bay (1982)	0.22 ± 0.034	0.78
Hudson River Plume	0.25 ± 0.021	0.94
Puget Sound		
1966	0.16 ± 0.026	0.60
1967	0.15±0.026	0.79
Narragansett Bay and MERL		
Microcosms (Keller, 1988*b*)	0.15 ± 0.004	0.82

Data modified from a tabular summary by Cole and Cloern (1987) and other sources as indicated. Productivity (π) was regressed on the composite parameter $C_s E_0/K$. The slope of the regression has units mol^{-1} photon.

Mechanistic Descriptions of the Relationship between Primary Production and Chlorophyll a

For remote sensing application, is desirable to relate π to the near-surface chlorophyll *a* concentration (C_s), which is easily measurable, rather than to the integral chlorophyll *a* concentration (C_i), which is not. Strictly, the chlorophyll *a* concentration seen by an airborne or satellite sensor corresponds to a weighted average value of chlorophyll a within the first attenuation depth (equal to 1/4.6 = 0.22 times the euphotic zone depth) and designated C_k. For most practical applications, it is reasonable to assume that $C_s = C_k$. Note that under the assumption of a uniform vertical chlorophyll a distribution $C_k = C_s = C_i/D = C_i/4.6K$.

Talling (1957 *a, b*) provided the original formulation for the dependence of π on chlorophyll a concentration, incident irradiance, vertical attenuation of scalar irradiance (K) and the Pl curve parameters P_m^{chl} and I_k:

$$\pi = C_s P_m^{abl} \log_e (2E_0/I_k)K \qquad ....(9)$$

This formulation has been verified in a range of freshwater environments, and a similar equation has been developed by Bannister (1974). Note the similarity of Talling's (1957*a,b*) model (Eq. 8) to a recent treatment of the same problem by Platt (1986):

$$\pi = C_s P_m^{chl} \log_e [E/I_k + \sqrt{1+(E/I_k)^2}]/K \qquad ... (10)$$

The difference between the equations of Talling (1957 *a,b*) and Platt (1986) arise from different assumptions about the shape of the PI curve. It can be easily demonstrated that Platt's (1986) model and Talling's (1957 *a,b*) model converge for large values of E_0/I_k.

The depth of the euphotic zone is not explicit in equations relating π to C_s and the surface irradiance does not enter as a linear multiplicative terms in Eqs. 9 and 10, but rather appears within the logarithmic term at the extreme right-hand side of these equations. How can the different formulations of the dependence of π on E_0 given in Eqs. 9 and 10 be reconciled with the linear dependence implied in Eq. 7? Nothing that $P_m^{chl} = \alpha^{chl} I_k = \phi_m a^* I_k$, and multiplying Eq. 9 bu E/E (= unity) leads to

$$\pi = [C_s a^*/K][\phi_m I_k \log_e (2E_0/I_k]E_0 \qquad ...(11)$$

The terms within the first set of square·brackets give the proportion of incident irradiance absorbed by phytoplankton, whereas the terms within the second set of square brackets give a measure of the efficiency with which absorbed light is used for photosynthetic carbon fixation, which is equivalent to η in Eq. 7. The last term on the right-hand side of Eq. 10 is the incident irradiance (E_0). Thus, a logarithmic dependence on E_0 is implicit in Eq. 7. Several recent theoretical models treat phytoplankton growth rate as a function of irradiance, photon yield, and the absorption coefficient.

Models of primary production that are conceptually adentical to Eq. 11 can be derived from these models of phytoplankton growth. Estimates of primary production from chlorophyll *a* and incident irradiance are dependent on a number of environmental and physiological factors. The environmental factors involve the partitioning of absorbed photons between phytoplankton and other substances and can be parameterised as the ratio of light absorbed by pytoplankton to total light attenuation. This has two components, the total pigment content within the photic zone and the vertical distribution of pigments. The physiological factors involved included genotypic and phenotypic modifications of the photosynthesis-light response curve and can be parameterized by P_m^{chl}, α^{chl}, $\alpha^* \phi_m$, and $I_k = P_m^{chl}/a^{chl}$. The ability to estimate primary production from chlorophyll *a* will ultimately be limited by the variability of these parameters.

9

LIGHT RESPONSE CURVE

Instantaneous rates of photosynthesis may be controlled by external factors, including irradiance, temperature, and the concentrations of CO_2 and O_2, by the molecular composition, of the photosynthetic apparatus, including the concentrations or activities of pigments, catalysts of electron transport, and enzymes of CO_2 fixation, or by other factors, such as the concentration of metabolites, or the demands for photosynthetic products. Photosynthesis-factor response curves, such as the photosynthesis–irradiance, photosynthesis–CO_2, and photosynthesis–temperature relations describe the response of photosynthesis rates to external factors. In addition, the interpretation of these curves often provides important information on regulatory control of photosynthesis by the composition of the photosynthetic apparatus and other internal factors. Interpretation of resource-response relations rests on the basic principle formulated by Blackman (1905) that an abrupt break in the response curve indicates a switch in control of photosynthesis between two different external factors.

Often, these external factors have their dominant influence on different components of the photosynthetic apparatus. For example, the efficiency of utilization of irradiance often limits the initial slope of the photosynthesis irradiance (PI) curve, but temperature or CO_2 availability often determines the light-saturated rate. In other words, the initial slope is often determined by light harvesting and the maximum rate by carboxylation. Usually, the simple resource-response relation envisaged by Blackman or a form based on Michaelis Menten kinetics is not observed in part because the concentrations of CO_2 and O_2 or irradiance at the site of photosynthesis within a microalgal cell or macroalgal thallus differs from that in the external environment. In an attempt to deal with this problem plant physiologists have developed models based on a mechanistic description of transport and fixation of CO_2 and irradiance distribution within leaves.

Although gradients in irradiance and resistances to CO_2 diffusion may be particularly important in macroalgae models similar to those employed with terrestrial vascular plants have not been applied to the algae. The response of photosynthesis to irradiance can be considered over the short and long term. Instantaneous obervations of CO_2 or O_2 exchange made over the short term are often considered to be uncoupled from growth, whereas the cumulative rate of photosynthesis should be consistent with the measured growth rate. In practice, however, it is often difficult to separate photosynthesis from growth.

Many techniques require finite time intervals for carrying out photosynthesis measurements during which growth can occur. Although it is often taken for granted that algal growth rate can

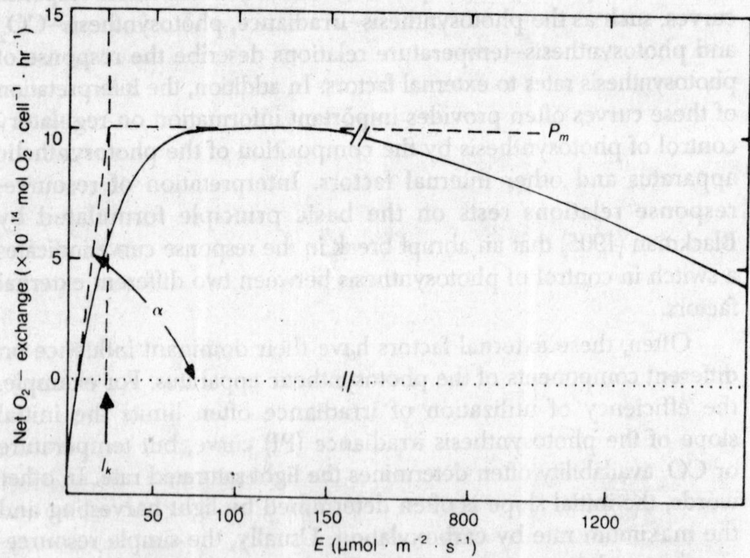

Fig. 9.1. Influence of irradiance on the photosynthetic rate of Chlamydomonas rheinhardtii previously grown at 15 μmol·m^{-2}·m^{-2}. P_m is the maximum rate of light saturated photosynthesis, α is the initial slope, and I_k (I_k=P_m/α) is the irradiance at the onset of light saturation.

be predicted from measurements of photosynthesis rate, it was not until 1979 that Bannister demonstrated that measurements of growth and photosynthesis in the green algae *Chlorella pyrenoidosa* were internally consistent. The difficulty in relating growth to photosynthesis arises largely from phenotypic, genotypic and ontogenetic changes in the size and composition of the photosynthetic apparatus with consequent changes in biomass- specific photosynthesis rates. Interpretation of the response of algal growth to irradiance has been given a mechanistic foundation in terms of the energetic efficiency and size of the photosynthetic apparatus.

The PI curve plays a central role in measuring, modeling, and predicting algal photosynthesis, and in assessing both intra and interspecific variations in photosynthetic physiology. It contains important information on the functioning of various components of the photosynthetic apparatus and their response to environmental variables. This chapter considers the equations that are used to describe the PI curve, the range of variability of the parameters of the PI curve, and problems encountered in measuring the PI curve.

Mathematical Descriptions of the Photosynthesis-Irradiance Curve

The relationship between net gas exchange and irradiance can be broken up into three regions; these are a light-limited region in which photosynthesis increases with increasing irradiance a light-saturated region in which photosynthesis is independent of irradiance and a photoinhibited region in which photosynthesis decreases with further increases in irradiance. In addition some workers have identified an intermediate region, where there is either a gradual or abrupt transition from light-limited to light-saturated photosynthesis. The photosynthesis-light response (PI) curve is constructed by plotting the biomass-specific rate of net photosynthesis against irradiance. Biomass may be expressed in terms of a number of possible variables including chlorophyll a, total pigment content, organic carbon, organic nitrogen, or dry weight.

Mathematical descriptions or models of the PI curve are empirical in that they attempt to describe the influences of all of the factors that may affect photosynthesis rates in terms of the minimum number of parameters required to adequately describe the PI curve. Most models of the PI curve characterize the light-limited and light-saturated regions with three parameters : these are the initial slope (α), the dark respiration rate (R), taken as the extrapolation of the initial slope to darkness, and the light-saturated photosynthesis rate (P_m). Inclusion of the photoinhibited region requires additional coefficients in the mathematical formulation and complicates the interpretation of the meaning of the parameters. This section begins with a consideration of light-limited and light-saturated photosynthesis and then proceeds to an examination of the complete PI curve, including photoinhibition.

Table 9.1. Different formulations of the photosynthesis light response curve for gross photosynthesis in the absence of photoinhibition.

$P = \alpha E$	$E < P_m/\alpha$	(5.1)
$P = P_m$	$E > P_m/\alpha$	
$P = \dfrac{P_m \alpha E}{P_m + \alpha E}$		(5.2)
$P = \dfrac{P_m \alpha E}{[P_m^2 + (\alpha E)^2]^{0.5}}$		(5.3)
$P = \alpha E \exp(\alpha E/P_m e)$		(5.4)
$P = aE \exp(-\alpha E/P_m e)$	$E < P_m e/\alpha$	
$P = P_m$	$E > P_m e/\alpha$	(5.5)
$P = P_m [1-\exp(\alpha E/P_m)]$		(5.6)
$P = \alpha E - [cE - (\alpha E)^2]/4P_m$	$E < P_m e/\alpha$	(5.7)
$P = P_m$	$E > 2P_m/\alpha$	
$P = P_m \tanh(\alpha E/P_m)$		(5.8)

The quantitative description of the light-dependence of photosynthesis dates from Blackman's (1905) studies of limiting factors in plant production. Blackman considered the rate of plant production to be linearly dependent on the availability of a single limiting factor at low resource availability may not be as abrupt as postulated by Blackman kinetics, and a number of different formulations of the PI curve were proposed by plant physiologists, oceanographers, and limnologists. Although algal physiologists often take the shape of the PI curve as fixed, it may in fact be variable, and vascular plant physiologists often explicity employ a shape factor. Abrupt transitions in the PI curve are thought to result from extremely efficient regulatory controls over the activity of the primary carboxylating enzyme RUBISCO, although these may be obscured due to gradients in the internal light environment.

Variations in the relative importance of these biochemical and nonbiochemical factors could have important consequences for mathematical descriptions of photosynthetic responses to irradiance as the greatest imprecision in current models is often found in the transition region.

In analogy with the Michaelis-Menton description of enzyme kinetics, the rectangular hyperbola has been used to stimulate the photosynthesis-light response in ecological models. Unfortunately, this equation consistently yields a poor fit to experimental data, especially in the transition region from light limitation to light saturation, because the rectangular hyperbola does not allow a rapid enough transition from light-limited to light-saturated photosynthesis. A modified rectangular hyperbola provides a much better fit to the experimental data by using higher order terms in the denominator, but this equation has no greater mechanistic justification than any of a number of other possible formulations.

The most popular models currently employed to describe experimental PI curves are the hyperbolic tangent function and an exponential model. The hyperbolic tangent function was originally introduced as a purely empirical function, but has found theoretical justification in an analysis by Chalker (1980). The exponential function is mathematically identical to the cumulative one-hit Poisson distribution used to describe the relationship between flash yield and flash intensity.

Jassby and Platt (1976) reformulated eight models of gross photosynthesis, so that they were expressed in terms of the same two parameters. These parameters are α, defined here as the light-limited initial slope of the PI curve at low incident irradiances and P_m, the light saturated maximum photosynthesis rate. To account for negative values of net photosynthesis in darkness, the authors needed a third parameter, R, the dark respiration rate. The various formulations of the PI curve differ in the abruptness of the transition from light-limited to light-saturated photosynthesis. When P_m, α, and R are assumed to have fixed values, the Blackman formulation greatly over estimates photosynthesis and the rectangular hyperbola greatly underestimates photosynthesis at intermediate light levels relative to the other six models.

The accuracy with which the different models describe the PI curve has been considered for free-living and symbiotic microalgae and macroalgae. Working with ^{14}C-uptake data for natural assemblages of marine phytoplankton, Jassby and Platt (1976) concluded that the hyperbolic tangent function provided the best fit to the data. However, a reanalysis of their data with improved curve-fitting techniques indicated that most models (i.e., could not be distinguished on a criterion of goodness of fit, although Blackman kinetics and the Michaelis Menton equation were clearly inferior descriptions of the PI curve. Chalker (1981) concluded that the hyperbolic tangent function and the exponential function provided more accurate estimates of α and P_m in the reef-building corals than did Blackman kinetics or the rectangular hyperbola, consistent with Lederman and Tett's (1981) analysis for phytoplankton.

Coutinho and Zingmark (1987) reported that the hyperbolic tangent provided the best fit to PI curves of marine macroalgae based on both the criteria of goodness of fit and minimizing residuals. Surprisingly, a recent investigation of the PI curve in marine macroalgae found poor fit and low precision in estimates of α and P_m using the hyperbolic tangent function. Researchers involved in terrestrial plant photosynthesis have circumvented the problem of choosing the most appropriate model by introducing the concept of convexity, in which the shape of the PI curve is determined by a parameter, which, like P_m or α, needs to be independently assessed. In the first application of the concept of convexity to algae, Leverenz *et al.* demonstrated that the shape of the PI curve depended on the degree of photoinhibition in *Chlamydomonas reinhardtii.* The transition from light-limitation to light-saturation became less abrupt with increased photoinhibition.

Alternative forms of the Photosynthesis-Light Response Curve

Most investigators express photosynthesis rates on a unit chlorophyll *a* basis because chlorophyll *a* is a readily measured index of the size of the photosynthetic apparatus. However, some investigators prefer to express photosynthesis rates in terms of the optical cross section and turnover time of a photosynthetic unit (PSU) rather than in terms of the chlorophyll *a* specific maximum rate P_m^{chl} and the initial slope a^{chl}. The analysis of steady-state photosynthesis in terms of PSUs is derived from the treatment of O_2 flash yield versus flash irradiance curves.

$$Y/Y_m = 1 - exp\,(-\sigma^{PSU}E) \qquad \text{...(1)}$$

where Y is the yield of O_2 per flash, Ym is the maximum yield of O_2 in a saturating flash, σ^{PSU} is the 'optical' cross-section of a PSU and E is the flash irradiance (µmol photons m^{-2} $flash^{-1}$). In this section, we show that the two approaches are equivalent when a exponential model *i.e.,* is employed. Following Mauzerall and Greenbaum (1989), the optical cross section (σ^{PSU}) can be considered to equal the product of the maximum photon yield for O_2 evolution, the *in vivo* absorption coefficient of a chlorophyll *a* molecule and the size of the PSU:

$$\sigma^{PSU} = \phi_m a^*/N^{PSU} \quad ...(2)$$

where ϕ_m is the maximum photon yield (mol O_2 mol^{-1}, photons), a* is the *in vivo* absorption coefficient normalized to chlorophyll *a* and N^{PSU} is the amount of chlorophyll *a* in a PSU. Thus, σ^{PSU} is analogous to the initial slope of the PI curve when photosynthesis is normalized to PSUs rather than to chlorophyll *a*; specifically,

$$\sigma^{PSU} = \phi_m a^*_{PSU} = \Phi_m a^* N^{PSU} \quad ...(3)$$

where a_{PSU} is the specific absorption coefficient normalized to a photosynthetic unit.

Based on Mauzerall's treatment of the flash yield versus flash irradiance curve, Dubinsky *et.al* (1986) proposed the following form for a continuous light PI curve:

$$P/P_m = 1-\exp(-\sigma^{PSU} \tau E) \quad ...(4)$$

where σ^{PSU} is the optical cross section of a PSU, τ is the turnover time of a PSU in continuous light, and E is the irradiance (µmol. m^{-2}. s^{-1}). Nothing that by definition $P_m^{PSU} = 1/\tau$ and $\sigma^{PSU} = \alpha^{PSU}$ allows Eq. 10 to be rewritten as

$$P^{PSU} = P_m^{PSU} [1-\exp\ (a^{PSU}\ E/P_m^{PSU})] \quad ...(5)$$

This is mathematically analogous to Eq.4.6. By noting the identities.

$$P^{chl} \equiv P^{PSU} N^{PSU}$$

$$\alpha^{chl} \equiv \alpha^{PSU} N^{PSU}$$

$$P_m^{chl} \equiv P_m^{PSU} N^{PSU} \quad ...(6)$$

One advantage gained by expressing both the initial slope and lightsaturated rate of photosynthesis in terms of the number of photosynthetic units is the explicit recognition that α and P*m* are not necessarily functionally independent. The often-observed covariation of α and P_m can be most readily explained in terms of changes in the number of active photosynthetic units, all other factors remaining unchanged. When α and P_m vary independently, it becomes necessary to invoke changes in τ, a^*_{PSU} *or* ϕ_m.

Mechanistic Interpretation of the Photosynthesis-Light Response Curve

As described above, the parameters of the PI curve are strictly empirical coefficients. To gain a greater understanding of the environmental or physiologic factors that regulate photosynthesis, it is desirable to relate these parameters to more fundamental processes. For the remainder of this discussion, we assume that the PI curve can be described by the exponential function:

$$P^{chl} = P_m^{chl}\,[1-\exp\,(-\alpha^{chl}\,E/P_m^{chl})] \qquad ...(7)$$

Both the initial slope (α^{chl}) and the light-saturated rate (P_m^{chl}) have been described in terms of underlying biochemical and biophysical events. At low irradiance, the photosynthesis rate is proportional to the amount of light intercepted by the photosynthetic apparatus, and the proportionality constant is the initial slope α^{chl}. The initial slope can thus be equated to the product of a chlorophyll α-specific absorption coefficient and the maximum realized photon yield of photosynthesis (ϕ_m) :

$$\alpha^{chl} = a^*\phi_m \qquad ...(8)$$

Absorption coefficients are often defined per unit chlorophyll *a* but can also be normalized to units of per cell or per unit mass. Much of the variation in α^{chl} in actively growing microalgae appears to be due to variations in light absorption (a^*), rather than to variations of the photon yield (ϕ_m). However, the initial slope has been observed to decrease under adverse growth conditions due to a decline of ϕ_m and interspecific variations of ϕ_m have been reported. During phenotypic responses to irradiance or nutrient limitation, compensating changes in ϕ_m and a^* may lead to the situation where there is less variation in α^{chl} than in either ϕ_m or a^*. Typically, the photon yield is calculated from measurements of the initial slope and the absorption coefficient by rearranging Eq. (8).

Only in the experiments of Welschmeyer and Lorenzen (1981) and Cleveland and Perry (1987) has ϕ_m been obtained directly from simultaneous measurements of absorption and photosynthesis using an integrating sphere (i.e., from the initial slope of a plot of photosynthesis versus absorbed light). Whether measured directly or indirectly, all estimates of ϕ_m are subject to potential errors in the measurements of light absorption. In addition, variations in gas exchange rates due to induction phenomena changes in mitochondrial respiration between light and darkness and changes in gas exchange with irradiance due to photorespiration produce ambiguities that confound an accurate estimate of ϕ_m. With material acclimated to low irradiances or nutrient-deficient conditions, light-limited rates of photosynthesis may only be marginally greater than the light compensation point, making it difficult to estimate the initial slope accurately. These uncertainties are manifested by a continuing controversy over the absolute upper limit of ϕ_m.

According to the Z-scheme for photosynthesis, a minimum of eight photons are required per molecule of O_2 evolved, giving a maximum upper limit for ϕ_m of 0.125 mol O_2, mol^{-1}, photon. The value of 0.125 is an upper limit, which is unlikely to be realized in practice because of expected inefficiencies in photosynthetic electron transport. Although many investigators report values for ϕ_m <0.125, a number of investigations are inconsistent with the Z scheme in that values of ϕ_m>0.125 have been observed.

In view of recently expressed reservations about the operations of the two photosystems and the demonstration of the direct photoreduction of $NADP^+$ by PS2 additional experimental determinations of ϕ_m under well-controlled conditions that are free of the experimental artifacts are clearly warranted. Largely as a consequence of problems associated with transitory gas exchange phenomena, some workers have suggested that the maximum value for ϕ_m can only be estimated with actively growing cells in balanced growth. Measurements of ϕ_m based on balanced growth the technically more difficult than estimates based on gas exchange, because of the need to maintain stable conditions of biomass and irradiance over several division cycles (i.e., days to weeks).

It is therefore, not surprising that only a few studies have attempted to estimate ϕ_m from measurements of light absorption and growth. Unfortunately, even these measurements do not provide unequivocal estimate of ϕ_m. because of difficulties associated with the determination of metabolic costs associated with cell synthetic activity. A finite amount of time is required for transfer of an electron from H_2O to CO_2 and that time interval (τ) is referred to as the turnover time of a PSU. Light saturation of photosynthesis occurs when τ exceeds the time interval between successive excitations.

Photosynthetic unit size was originally defined in terms of the number of PS2 reaction centers, determined from oxygen flash yield experiments and, until recently, it was assumed that the ratio of PS2: cyt b_6/ϕ : PSI was fixed at unity, based on the Z scheme for photosynthetic electron flow. It is now known that the photosynthetic electron transport chain is less rigidly defined than is required by a literal interpretation of the Z scheme, and various stoichiometries of reaction centers and electron transport chain catalysts have been observed.

Table 9.2. Stiochiometry of electron transport chain components and rubisco in microalgae.

Species	*Irradiance µmol.m-2.s-1*	*RCII*	:	*cytblf*	:	*RCI*	*Rubisco*
Chlamydomonas	47	1.3		1.4		1.0	–
reinhardtii	200	2.2		–		1.0	–
	400	2.6		2.8		1.0	–
Chlorella	low	1.4		–		1.0	–
pyrenoidosa	high	1.3		–		1.0	–
Dunaliella tertiolecta	45	0.9		–		1.0	–
	600	0.8		–		1.0	–
Duralielia tertiolecta	80-1,900	0.8		1.4		1.0	0.9-3.5
Mantoniella	5	3.7		1.2		1.0	–
squamata	100	2.4		1.8		1.0	–
Isochrysis galbana	30-600	2.0				1.0	–
Isochrysis galbana	(N-limitation)	1.2		4.0		1.0	5.3-14.4
Skeletonema	30	2.3		–		1.0	–
costatum	200	1.4		–		1.0	–
	600	1.1		–		1.0	–
Ny 17	27	0.89		–		1.0	–
	274	0.76		–		1.0	–
UP 45	27	0.44		–		1.0	–
	274	0.52		–		1.0	–
Skel	27	0.85		–		1.0	–
	274	0.63		–		1.0	–
Thalassiosira weisflogii	30-600	2.0		–		1.0	–
Prorocentrummicans	70-600	1.4		–		1.0	–

When dealing with intact cells, the photosynthetic unit can be empirically defined as all of the catalysts or compounds associated with a PS2 reaction center that are required for whole-chain electron flow from H_2O to CO_2, resulting in the evolution of one O_2 molecule. Thus, P_m^{chl} can be expressed as :

$$P_m^{chl} = 1/\tau N^{PSU} \quad ...(9)$$

One disadvantage of describing whole-chain electron flow in terms of photosynthetic unit number and turnover time of an entire PSU is that this approach avoids the question of what process actually sets the limit on PSU turnover during steady photosynthesis in continuous light. Light-saturated photosynthesis, as determined by net gas exchange techniques, may be set by reactions downstream and to some extent independent of photosynthetic electron transport. Extending the concept of the photosynthetic unit to include the catalysts of these downstream reactions would greatly dilute the original intent in defining the PSU in terms of a component of photosynthetic light harvesting and electron transport. Use of 18_O techniques to obtain true rates of gross oxygen evolution should lead to an increased understanding of the factors that regulate light-saturated photosynthesis. Significant rates of light-dependent oxygen consumption at light saturation may indicate an excess of electron transport capacity over carboxylation capacity.

Photosynthesis at Irradiances near the Light Compensation Point

The light compensation point is that irradiance at which net gas exchange is zero because gross photosynthesis and respiration rates are in balance. The compensation irradiance can be related to the initial slope and respiration rate as follows:

$$E_0 = R^{chl}/\alpha^{chl} \qquad ...(10)$$

where E_0 is the compensation irradiance (μmol photons. m^{-2}, s^{-1}); R^{chl} is the respiration rate (μmol O_2. mg chl $\alpha^{-1}.s^{-1}$) and, α^{chl} is the initial slope of the PI curve based on chlorophyll content (mol O_2. mol^{-1}, mg^{-1} chl*a*). The compensation irradiance is not fixed but varies primarily due to R^{chl} and to a lesser extent α^{chl}, both of which will depend on the environmental conditions (irradiance, nutrient availability, temperature), under which the algae were growing. A careful examination of photosynthesis rates at irradiances near the compensation point often reveals a break in the initial slope, in which the slope abruptly increases at low irradiances. The change in slope is referred to as the Kok-effect in recognition of the contributions of the investigator who first studied this phenomenon.

The Kok effect can be overlooked if the PI curve is poorly resolved near the compensation irradiance and is ignored in all mathematical treatments of the PI curve. However, the Kok effect may provide considerable insight into the relationship between

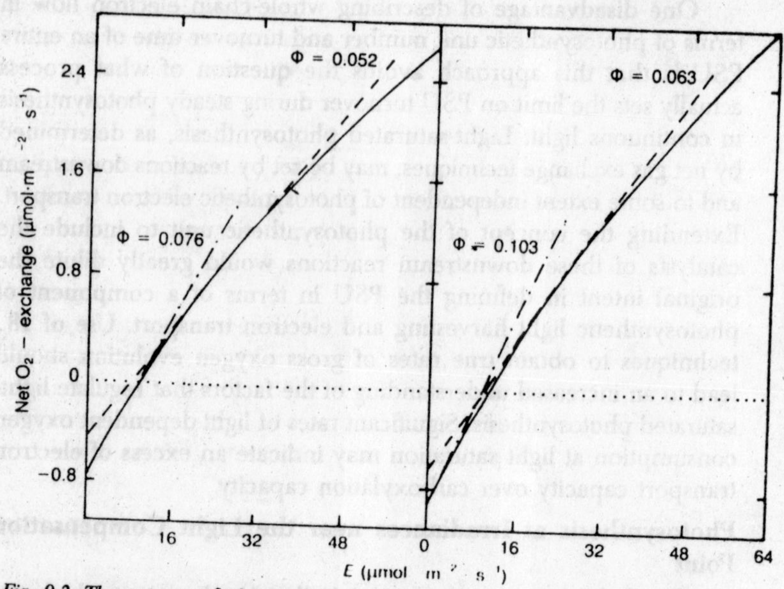

Fig. 9.2. The response of photosynthesis to limiting irradiance in the cyanobacterium Anacystis nidulans previously grown at 200 μmol·m^{-2}·s^{-1} to indicate the influence of Kok effect on the photon yield (ϕ_m) of photosynthesis in the absence (a) or presence (b) of nitrate. Figures next to the dashed line in (a) and (b) represent the calculated slopes of the linear least-squares regression analysis of the photon yield above (---) and (_.) below the light compensation point.

photosynthetic and respiratory metabolism. Kok (1948) found that the ratio of the slopes below and above the break in *Chlorella* varied with physiological condition by up to a factor to two; in the extreme case, an extrapolation of the upper line to zero irradiance gave an estimate of respiration that was only half that of the observed dark respiration rate. Kok (1948) initially interpreted the change in slope as a reflection of changes in the photon yield of photosynthesis (ϕ_m). However, it is now thought that the higher slope below the compensation point reflects a partial light inhibition of the respiration rate at low irradiances.

The Kok effect is not always observed, suggesting that light-inhibition of respiration may be transitory or only occur under certain physiological conditions. It has also been suggested that the efficiency of metabolite transport across the chloroplast envelope and the rate of cytosolic ATP consumption may determine the magnitude of the Kok effect. Osborne and Geider (1987a) hypothesized that the Kok effect would be particularly evident following addition of nitrate to nitrogen starved cells, as has recently been confirmed for the cyanobacterium *Anacystis nidulans.*

Models that include Photoinhibition

Reductions of biomass-specific photosynthesis rates in samples incubated at or near the surface are often observed under in situ or simulated *in situ* conditions. Equations that account for this high irradiance inhibition of photosynthesis have been developed by several investigators working in both fresh and salt waters. Platt *et al.* (1980) extended the exponential formulation.

$$P=P_s[1-\exp(-aE)]\exp(-bE) \quad ...(11)$$

Where $a = \alpha/P_s$ and $b = \beta/P_s$. In this formulation P*s* is not necessarily equivalent to the P_m of Eq. When photoinhibition is not observed $P_m = P_s$. However, when photoinhibition is observe $P_s > P_m$. Furthermore, the value of a determined from just the nonphotoinhibited portion of the PI curve will equal the sum $\alpha + \beta$ determined from a complete PI curve that includes photoinhibition. The three-parameter model of gross photosynthesis works well in many environments, but fails to adequately describe the extended plateau observed in some samples. Coutinho and Zingmark (1987) report that Eq. (11) provides a good fit to those PI curves of marine macrophytes that exhibit photoinhibition.

Fasham and Platt (1983) derived a four-parameter model of gross photosynthesis, including photoinhibition, which allowed more flexibility than previous models. Although derived from a description of the light reactions of photosynthesis that are typified by time constants of less than a second, the model was applied to data obtained from 4 to 6 hour incubations. This mismatch in time scales between the mechanistic processes incorporated into the model and the observations to which the model was applied may limit the usefulness of this model in other situations.

Megard *et al.* (1984) employed a three-parameter model that extended the rectangular hyperbola to include photoinhibition through addition of an E^2 term in the denominator,

$$P=P_s/(K_1 + E + E^2/K^2) \quad (12)$$

This equation is identical to Haldane's (1930) formulation for enzyme reactions that are inhibited by high substrate concentrations. A mechanistic derivation involving light dependent activation and inactivation of PS2 was provided by Megard *et al.* (1984), and a somewhat formulation of the same model was proposed by Peeters and Eilers (1978).

Reduced gas exchange rates at high irradiance are due largely to two processes: photoinhibition and potorespiration. Photoinhibition is a time-dependent process that appears to involve a reversible or irreversible inactivation of PS2. Photorespiration is caused by the oxygenase reaction of ribulose bisphosphate carboxylase-oxygenase, which is increased under high light and / or reduced CO_2 supply or increase O_2 concentration. An additional oxygen-consuming reaction in illuminated plants, the Mehler reaction, may play a role in the protection of the photosynthetic apparatus from photoinhibition.

High light inhibition of photosynthesis occurs on a time scale that is commensurate with the time required to determine the PI curve. It is thus likely that models that characterize high light inhibition of photosynthesis by a single, time invariant parameter may not adequately describe the light dependence of photosynthesis on shorter time scales. This question has been addressed by Gallegos and Platt (1985) and Neale (1987).

Interspecific and Intraspecific Variations in P_m and α

Photosynthesis rates are normalized to some measure of biomass to allow comparisons between organisms with widely different sizes and thus to account for the general observation that the rate of photosynthesis seales with plant biomass. The choice of a biomass unit for normalizing photosynthesis rates is often based on convenience, but this choice is important in making interspecific and intraspecific comparisons. In microalgae, photosynthesis rates are often expressed as cell specific or chlorophyll-specific rates or more rarely as carbon-specific or mass-specific rates. In macroalgae, photosynthesis is often normalized to unit area, dry weight or chlorophyll, and occassionally to unit wet weight or carbon content. The PI curve will have the same shape whatever the unit of biomass to which photosynthesis rates are normalized. However, changes in the absolute values of biomass-specific photosynthesis rates in response to environmental factors will depend on the unit of biomass to which photosynthesis rates have been normalized.

One goal of algal photosynthesis research is to determine the magnitude of phenotypic and genotypic variations in the parameters of the photosynthesis-light response curve. Phenotypic variation can be examined by comparing the attributes of genetically identical organisms grown under a range of environmental conditions, whereas genotypic variation is investigated by comparing photosynthesis and biomass of a range of genetically unique organisms grown under identical and well-defined conditions. Phenotypic responses to irradiance have been best characterized, whereas responses to light quality, nutrient availability, and temperature are not well understood. The effects of both phenotypic and genotypic variability of the photosynthetic apparatus of algae have rarely been examined in the same investigation, although the work of Gallagher et al. (1984) is a notable exception. The remainder of this section attempts to summarize our current understanding of the influence of phenotypic and genotypic variations on photosynthetic physiology.

Photoadaptation of Photosynthesis

An examination of the response of the photosynthetic apparatus to irradiance, nutrient availability, or temperature requires that all other factors be controlled. Photosynthetic parameters can vary widely in batch cultures of marine microalgae demonstrating the importance of ensuring that cells are in balanced growth before attempting to examine photoadaptation. However, it may take 5-20 generations before phytoplankton become fully acclimated to the prevailing environmental conditions. It is arguable that balanced growth is never achieved in batch cultures, because the organisms are continuously modifying their environment, which will in turn modify growth processes. Turbidostats can be used to maintain cells in light-limited balanced growth and chemostats can be used to maintain cells in nutrient-limited balanced growth. However, it is possible to maintain cells in nearly balanced growth in both nutrient-sufficient and nutrient-limited cultures by frequent subculturing.

Algae commonly respond to reduced irradiance by increasing the quantity of photosynthetic pigments. This can be viewed as an adaptive response in which the components of the light harvesting apparatus increase in abundance when light is a limiting resource. In nutrient-sufficient microalgae, a plot of the carbon-to-chlorophyll *a* ratio against irradiance is often linear, although the relationship is sensitive to temperature, with higher carbon to-chlorophyll *a* ratios at lower temperatures. In addition to changes in the size of the photosynthetic apparatus, as indicated by lower carbon-to-chlorophyll *a* ratios, the response to irradiance may involve changes in the composition of the photosynthetic apparatus.

The most common response to low irradiance is an increase in the antennae size of the photosynthetic unit, which is reflected in an increase in the ratio of total chlorophyll *a* to reaction centre

P_{700}. The classic pattern for the acclimation of the PI curve to irradiance is Myer's (1970) data for *Chlorella pyrenoidosa.* To understand this response, it is necessary to consider both chlorophyll *a*-specific and dryweight-specific photosynthesis rats. Myers (1970) found that the chlorophyll *a*-specific initial slope of the PI curve was largely independent of the irradiance at which the cells were grown but the chlorophyll *a*-specific light-saturated rate (P_m^{chl}) decreased at low irradiance.

In contrast, the dry-weight-specific light-saturated photosynthesis rates (Pm^w) was less dependent on irradiance, but the initial slope (α^w) decreased at low irradiance. Myer's (1970) observations suggest that light limited photosynthesis is controlled by the efficiency of light absorption, which convaries with pigment concentration. What controls light-saturated photosynthesis is less clear, but it appears to be the concentration of some component(s) which covary(ies) with algal mass. Foy and Gibson (1982b) found that cell- protein-specific P_m was independent of irradiance in *Oscillatoria agardhii*, whereas dry-weight-specific P_m decreased at high irradiance, presumably because of the accumulation of noncatalytic carbohydrate storage reserves.

It is not surprising that cell protein provides a good predictor of light -saturated photosynthesis given the large contribution of RUBISCO and thylakoid proteins to total cell protein. Other patterns in the response of chlorophyll *a*-specific and weight-specific photosynthesis rates have also been described. Richardson et al. (1983), for example, list five different types of response of the PI curve to preconditioning irradiance. It is possible, however, that the range of responses reported by Richardson *et al.* (1983) is not an inherent property of different algal groups but represents different components of a continum of photoadaptation along a gradient from light-limitation to photoinhibition.

Chlorophyll-specific, light-saturated photosynthesis rates generally increase in microalgae acclimated to high irradiance. Two-to five fold changes of (P_m^{chl}) in response to defferences in irradiance have been documented in a wide range of microalgae, including chlorophytes, diatoms, dinoflagellates, cyanobacteria and a rhodophyte. An almost linear increase of P_m^{chl} with the logarithm of irradiance is observed in several diatoms grown under nutrient-sufficient conditions in continuous light. In contrast to P_m^{chl}, α^{chl} generally varies little with preconditioning growth irradiance. For

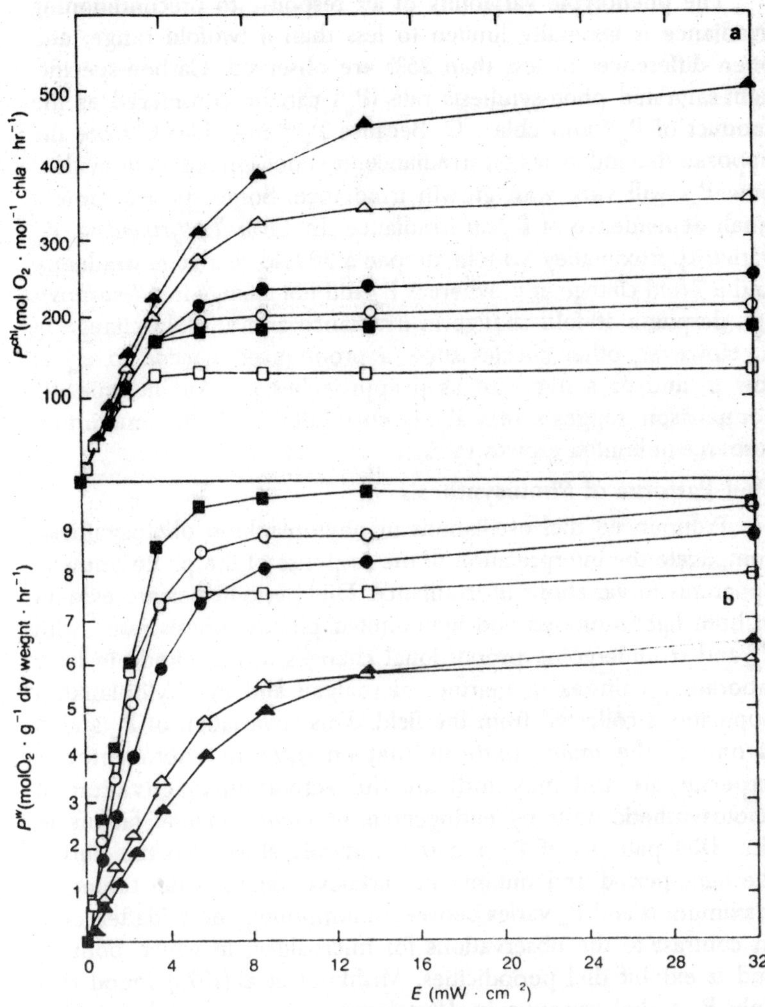

Fig. 9.3. Response of photosynthesis-irradiance (PI) curve to light-limited growth in Chlorella pyrenoidosa.

example, Foy and Gibson (1982a) found small changes in α^{chl} in response to an increase in growth irradiance for 30 to 150 μmol. m^{-2}.s^{-1} in 20 strains of freshwater cyanobacteria. The value of α^{chl} increased upto 20% in 5 strains, decreased upto 25% in 15 strains, with a mean overall decrease of 6% where data from all 20 strains are considered.

The phenotypic variability of a^{chl} response to preconditioning irradiance is generally limited to less than a twofold range, and often differences of less than 25% are observed. Carbon-specific, light-saturated photosynthesis rate (P_m^c) can be considered as the product of P_m^{chl} and chla : C. Because P_m^{chl} and chla: C have the opposite dependencies on irradiance, it is not immediately evident how $P_m c$ will vary with growth irradiance. Some species show a small dependence of P_m^c on irradiance. In *Chlorella pyrenoidosa*, P_m^c varied approximatley 1.5-fold, despite a 20-fold change in irradiance and a 7-fold change in µ, whereas P_m^c did not change in *Glenodinium* sp., despite a 10-fold change in irradiance and a 3-fold change in µ. However, other species show a pronounced decline in P_m^c at low µ, and/or a decrease as µ approaches μ_m. An interspecific comparison suggests that P_m^c is correlated with the maximum resource-unlimited growth rate.

Diel Patterns of Photosynthesis

Pronounced diel oscillations in phytoplankton photosynthesis complicate the interpretation of the response of the photosynthetic apparatus to variations in irradiance. Diel periodicities are evident in both light-saturated and light-limited photosynthesis rates, with P_m and α undergoing proportional changes in magnitude in both laboratory cultures of marine microalgae and in phytoplankton populations collected from the field. This covariation of P_m and α is one of the major patterns that emerges in photosynthesis experiments and may indicate the activation/inactivation of photosynthetic units by endogenous or environmental factors in situ. Diel patterns of P_m and α commonly show maxima during the light period and minima in darkness, although the timing of maximum α and P_m varies between midmorning and midafternoon. In contrast to the observations for microalgae, in which both P_m and α exhibit diel periodicities, Mishkind et al.(1979) found that only P_m a diel variation in *Ulva lactuca*, whereas α did not. The diel variation in P_m was attributed to an endogenous rhythm in the rate of reoxidation of plastoquinone by PSI.

In a comprehensive examination of 24 marine phytoplankton species grown at 15°C and 1,200 µW.m^{-2} (approximately 60 µmol. m^{-2}.s^{-1}) on a 12:12 light-dark cycle, Harding *et al.* (1981) documented diel variations in 17 species. A 150-fold difference was observed for P_m in the diatom *Ditylum brightwellii*, and a 50-60-fold difference was observed in *Lauderia borealis*. Of the species that exhibited no diel periodicity, most were small and fast growing and these included three common laboratory "weeds" (i.e., the diatoms *Phaeodactylum tricornutum* and *Skeletonema costatum* and the chlorophyte *Dunaliella tertiolecta*). In a subsequent investigation using. *D. brightwellii.* Harding *et al.* (1981) showed that diel periodicity in the PI curve was most pronounced in early exponential phase cultures where cell-specific maximum photosynthesis rates (P_m^{cell}) varied from < 10 to 170 pg $C.cell^{-1}.h^{-1}$, and was least evident in late stationary phase cultures, in which P_m^{cell} varied from 4.5 to 7.5 pg $C.cell^{-1}.h^{-1}$.

Nutrient Dependence of Photosynthesis

The components of the photosynthetic apparatus, namely the pigmentprotein complexes, the carriers of the photosynthetic electron transport chain, and enzymes of the photosynthetic carbon reduction cycle, account for a large fraction of total cell nitrogen in micro- and macroalgae. For example, Falkowski *et al.* (1989) calculated that 20 to 30% of total cell nitrogen could be attributed to the carboxylating enzyme RUBISCO in nutrient-sufficient *Isochrysis galbana.* The nitrogen costs of the light-harvesting complexes were determined by Raven (1984) to range from 28 mol nitrogen, mol^{-1} chromophore in the chlorophyll *a* and *b* complexes of the chlorophyta to 197 mol nitrogen. mol^{-1} chromophore associated with allophycocyanin in the phycobilliphytes. It is therefore not surprising that algae respond to a reduction in nitrogen availability by reducing the size of the photosynthetic apparatus. This reduction in pigmentation may be adaptive, if it allows a redistribution of nitrogen from light harvesting proteins to other catalysts or if it affords greater protection from photoinhibition when growth processes limit the rate of consumption of the products of photosynthesis. however, a reduction in the chlorophyll *a*; carbon ration (chlorosis) appears to be a general response to nutrient limitation, and is not specific to nitrogen limitation and macroalgae.

The range of chla : C encountered is likely to depend on both irradiance and temperature, with low temperature and high irradiance contributing to decreases of chal : C, which are most pronounced at low nutrient-limited growth rates. In addition to a role as light-harvesting accessory pigments, phycobiliproteins have been suggested to be a major intracellular nitrogen reserve in cyanobacteria (Boussiba and Richmond, 1980) and red algae. Although efficiently coupled to photosynthesis in low light environments, phycobiliproteins may become uncoupled from photosynthesis under high irradiances. Phycobiliproteins decrease to a greater extent than chlorophyll *a* under nitrogen-limited conditions with much of the nitrogen in the phycobiliproteins presumably being mobilized to allow continued growth in a nitrogen-limited environment.

In contrast to the phycobiliphytes in which the accessory pigment-to-chlorophyll *a* ratio decreases under nutrient-limited conditions, both chlorophyll c: chlorophyll *a* and carotenoid: chlorophyll *a* increased with increasing N-limitation in *Isochrysis galbana.* Chlorophyll *a*-specific, light-saturated photosynthesis rates typically decline slightly under moderate nutrient-limitation, dropping off markedly only as μ approaches zero. A linear dependence of carbon-specific, light-saturated photosynthesis rates

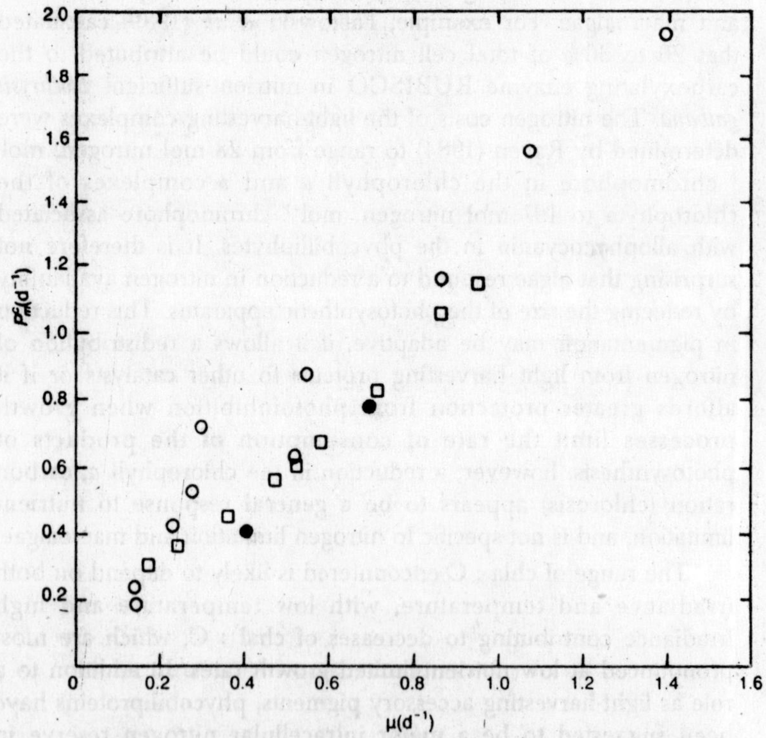

Fig. 9.4. Variation in carbon-specific, light-saturated photosynthesis (P_m, with unit of d^{-1}) as a function of growth rate (μ, in unit of d^{-1}) for nutrient-limited phytoplankton.

on nutrient-limited growth rate has been observed in several microalgae and a macroalga.

Correlated with the reduction in P_m^c is a decrease in the proportion of cell nitrogen, which is associated with RUBISCO. This response to nutrient limitation involves large reductions in chlorophyll per cell and slight reductions in the chlorophyll-specific light-saturated photosynthesis rate. In general, chlorophyll *a*-specific rates of light-limited photosynthesis are less affected by nutrient limitation than are light-saturated rates. The initial slop was independent of growth rate in phosphorus-limited chemostat cultures of *Scenedesmus quadricauda* and phosphorus starved batch cultures of *Anabaena wisconsinense* and *Chlorella pyrenoidosa.* However, α^{chl} increased slightly with increasing nitrate-limitation in *Phaeodactylum tricornutum* and *Isochrysis galbana.* A large increase in the chlorophyll *a*-specific absorption coefficent was partially offset by decreases in the maximum photon yield of photosynthesis in these microalgae. Osborne and Geider (1986) observed that α^{chl} increased form 0.0045 to 0.0065 $m^2.mg^{-1}$, chl*a*, but ϕ_m decreased from 0.129 to 0.10 mol O_2 mol^{-1} photons.

Herzig and Falkowski (1989) observed an increase in α^{chl} from 0.0094 to 0.016 $m^2.mg^{-1}$ chl*a*, and a decrease in ϕ_m from 0.08 to 0.06 mol O_2 mol^{-1} photons. In contrast to these observations for nitrate-limited cultures, photon yields were approximately 3-4 times lower in vitamin B_{12}-limited *Monochrysis lutheri* than in nutrient-sufficient cultures. More dramatic declines of ϕ_m and α^{chl} have been observed in stationary phase batch cultures, indicating a large reduction in photosynthetic efficiency under nitrogen-starved conditions.

Temperature Dependence of Photosynthesis

Most investigations of the temperatrue dependence of photosynthesis have examined the response of photosynthesis rates or enzyme activities immediately following a shift in temperatrue from the growth temperature to a higher or lower experimental temperature. A theoretical framework for interpreting these transient responses is provided by the Arrhenius plot in which the natural logarithm of the rate constant of interest is plotted against the inverse of the absolute temperature. The linear portion of an Arrhenius plot fits the equation.

$$\log_e k = -(E_a/RT) + C \qquad ...(13)$$

where k is the rate constant (s^{-1}). E*a* is the activation energy ($kJ°/mol^{-1}$), R is the universal gas constant (8.3 mol^{-1}, k^{-1}). T is the absolute temperature (°K) and C is an arbitrary constant. Plots of In k versus 1/T are linear over a range of temperatures near the growth temperature but may show a change in slope within this range. Outside of this temperature range, k declines precipitously. Arrhenius constants for light-saturated photosynthesis and for various partial reactions have been summarized by Raven and Geider (1989). Light-saturated photosynthesis and the carboxylase activity of RUBISCO are characterized by an E_a of 54-72 $kJ.mol^{-1}$, which corresponds to a Q_{10} of 2.1–2.7. This is slightly greater than the Q_{10} for maximum (i.e., resource unlimited) growth (μ_m) of microalgae obtained from interspecific comparisons, but lower than the Q_{10}, which characterizes the phenotypic response of μ_m to temperature . When considering a seasonal cycle, temperature is the environmental factor that consistently accounts for the largest component of the variance in P_m^{chl} in marine phytoplankton assemblages with 50% of the variation in P_m^{chl} accounted for by variations in temperature.

Recent investigations demonstrate that α^{chl} shows a similar temperature dependence to P_m^{chl}, immediately following a shift in temperature. This contrasts with the commonly held expectation that light-harvesting and light-limited photosynthesis should be independent of temperature. The temperature dependence of α^{chl} is not limited to short periods following a change in temperature but can be seen in samples from the natural environment, as well. For example, Cote and Platt (1983) report that temperature can account for 56% of the variance in α^{chl} within the temperature range 5-16°C in Bedford Basin, Canada. this is similar to the proportion of variance in P_m^{chl}, which can be attributed to temperature.

Reductions in α^{chl} at low temperature indicate that the maximum photon yield of photosynthesis (ϕ_m) may be temperature dependent, although it is not possible to confirm this suggestion without independent measurements of the chlorophyll *a*-specific light absorption coefficient (a^*). It is now well established that ϕ_m in terrestrial C_3 species is temperature sensitive because of the differential sensitivity of the carboxylase and oxygenase activities oı the RUBISCO enzyme to variations in temperature . Observations that the light-limited growth rate is often reduced at low temperatures to a greater extent than would be expected on the basis of a reduction in pigmentation supports the suggestion that ϕ_m is reduced at low temperature. However, this is not consistent with an effect on RUBISCO kinetics as the oxygenase activity would be predicted to decrease at low temperatures resulting in an increase in ϕ_m. Other evidence indicating an effect of temperature on ϕ_m is the temperature dependence of growth yields (dry weight of biomass produced per mole of photons absorbed) of *Chlorella vulgaris* and *Chlorogonium* sp., although in the case of these organisms the temperature effect was attributed to changes in bulk biochemical composition and respiratory metabolism. There is a clear need for examining the temperature dependence of the photosynthetic energy conversion efficiency and algal growth, because many natural environments are characterized by temperatures considerbly lower than the 15-25°C commonly employed in the laboratory.

Acclimation of an organism to a new temperature may modify the properties of catalysts and membranes, or change the relative amounts of different components and so influence photosynthesis rates in ways not predicted by short-term responses to temperature. The most striking acclimation to different temperatures involves reductions in chla : C and chla cell-1 at low temperatures and increases in the activity of photosynthesis rates at an assay temperature of 15°C were linearly related to *in vitro* RIBISCO activities determined at a temperature of 20°C in *Laminaria saccharina* acclimated to temperatures between 0 and 20°C. He found identical activation energies (i.e., the slope of Arrhenius plot) and temperature dependencies of half saturation constants for TCO_2 uptake in plants previously acclimated to 5°C and 20°C.

Genetic Variability

Phenotypic variations in photosynthetic physiology and growth rate in response to environmental conditions were considered in the preceding sections. These phenotypic responses have often formed the basis for modeling growth of aigal populations or assemblages. Good agreement, for instance, was found between the growth rate of natural populations of *Skeletonema costatum* in cage cultures and predictions based on the response of axenic cultures of a single strain of *S. costatum* to irradiance, temperature, and silicate concentration in the laboratory. However, few rigorous comparisons of predictions based on laboratory experiments and measurements of growth or photosynthesis in nature have been undertaken, and the validity of extrapolating from observations on a relatively few species maintained in the laboratory to mixed assemblages growing under natural conditions has not been firmly established. One factor that might limit this autoecological approach is genetic variability, both between different clones of morphologically indistinguishable organisms and between different taxonomic species. Here we examine the magnitude of intraspecific (clonal) and interspecific genetic variability of algal responses to different environmental conditions. Unfortunately, despite the importance of assessing clonal and interspecific variations in algal physiology, there have been few investigations of the effects of genetic variation on algal photosynthesis. It has been necessary to supplement data on instantaneous photosynthetic responses with information on algal in order to address this problem more comprehensively.

Clonal Variability

Clonal variability in the growth rate of various marine phytoplankters has been most extensively examined in a series of investiagations undertaken by Brand. Brand (1981) found statistically significant differences in the growth rates of different clones isolated from the same water mass, and significant differences in population growth rates from different water masses. Different populations were characterized by coefficients of variation in reproduction rate, which ranged from 3 to 12% under controlled experimental conditions. Consistent with these observations, Krawiec (1982) found little clonal variability in maximum reproduction rates and temperature tolerance of six clones of *Thalassiosira rotula* isolated from Narragansett Bay.

A single isolate from Baja, California had a significantly higher maximum division rate and a different temeprature tolerance than the Narragansett Bay isolates. However, in both the Baja clone and a Narragansett Bay clone the assimilation number (P^{chl}) showed a similar dependence on temperature and irradiance. Gallagher (1982) found much higher variability in reproduction rates of *Skeletonema costatum* isolated from Narragansett Bay, with the clones under investigation having a coefficient of variation of reproduction rate of approximately 70%. However, Gallagher's (1982) cultures were not maintained in balanced growth, and the assessment of the significance of the observed clonal variability is difficult to evaluate. Although variations of up to 10% in growth rate within a population are large by the yardstick of a geneticist, they are relatively small when compared with phenotypic variability. Phenotypic and genotypic components of variability in the photosynthetic apparatus of *Skeletonema costatum* were investigated by Gallagher *et al.* (1984). Three clones that varied in maximum specific growth rate and electrophoretic mobility patterns were cultured under conditions of nutrient-sufficient balanced growth at both high and low irradiance.

Significant genotypic and/or phenotypic differences were observed in a number of variables. However, phenotypic variability of P_m^{chl} and α^{chl} were independent of irradiance in one clone, but were positively related to irradiance in a second clone and inversely related to irradiance in a third. However, covariation of αchl and P_m^{chl} in these clones of *S. costatum* may indicate that changes in the photosynthetic apparatus are constrained by similar underlying physiological processes. A similar covariation of α_{cell} and P_m^{cell} was found in two *Synechococcus* clones cultured under a range of conditions.

Ecotypic and development differences in the photosynthetic charactersitics of *Laminaria saccharina* were investigated by Gerard (1988) with plants taken from shallow and deep-water sites in Marine and a turbid site in New York State. Juvenile *L. saccharina* were collected from the field sites and grown 6 weeks at four different light levels (0.065, 0.012,0.26, and 0.54 of the incident irradiance), with nitrate and phosphate added at 2-*d* intervals. Gerard (1988) found that area-based measurements of P_m and α were proportional to chlorophyll *a* content, but that plants varied in chlorohyll content, such that the turbid-water ecotype outperformed both the shallow-water and the deep-water forms at all irradiances. Plants from the deep-water site had lower specific growth rates under both light-saturated and light-limited conditions than plants from highest growth rates at radiances. Significantly, P_m^{chl}, α^{chl}, and the photon yield of photosynthesis were similar in all three ecotypes. Thus, ecotypic variation in *Laminaria saccharina* was due primarily to differences in pigment content rather than to fundamental differences in the cholorophyll specific rate of photosynthesis.

Interspecific Variability

Examining interspecific variability in photosynthetic gas exchange rates is complicated by the phenotypic and clonal variability outlined above. However, when organisms are cultured under identical experimental conditions, it is possible to examine the variability in specific growth rates, photosynthesis rates, or biochemical composition, which can be attributed to genetic variation. When such comparisons are made, much of the interspecific variability can be accounted for by higher taxonomic affiliation and organism size in the mircoalgae and by taxonomic affiliation, size, and form in the macroalgae. However, there must be a careful consideration of the appropriate physiologic attributes used in any comparison. For example, Dunstan (1973) found that five species of chlorophyll *c*-containing microalgae could be characterized by identical values of I_k, which implies little interspecific variability, but that chlorophyll *a*-specific light-saturated photosynthesis and dark respiration rates varied by about a factor of two, implying rather more variability. This section begins by considering the magnitude of interspecific variability, before proceeding to a discussion of the extent to which interspecific variability is predictable.

Maximum, resource-unlimited growth rates (μ_m) of microalgae, growing under optimal conditions vary by over an order of magnitude. For example, at 20°C, μ_m can be expected to range from approximately 0.1 d^{-1} in some large slow-growing dinoflagellates to approximately 3 d^{-1} in some small fast growing diatoms and green algae. This 30-fold range may approximate an upper limit on the amount of interspecific variability in μ_m that will be observed under controlled conditions. In fact, the range in μ_m is less than fivefold in many interspecific comparisons, even when organisms have been deliberately chosen to include both fast and slow growing species. To make meaningful assessments of interspecific variability in growth and photosynthesis, it is necessary to restrict the scope of the comparison to well-defined higher taxonomic groups. Table 9.3, for example, examines the variability in growth rates within the cyanobacteria, diatoms, dinoflagellates, and greenalgae. For comparison, Table 9.4 examines the varibility of some indicators of the size (C : chla, chla: protein) or operating performance (P^{chl}, P_m^{chl}, α^{chl}) of the photosynthetic apparatus.

This information is limited to those studies in which four or more species or clones were cultured under relatively well-defined environmental conditions. Within each of the taxa under consideration, μ_m varied by up to a factor of five and coefficients of variation for μ_m varied by up to 140%. This interspecific variation is considerably greater than that observed for clonal variability in which coefficients of variation of approximately 10% are typically observed. Photosynthetic characteristics are considerbly less variable than μ_m, with coefficients of variation of 25 to 50% for P_m^{chl} and α^{chl} and 10 to 40% for the C ; chla and protein : chla ratios. Apparently exceptional in this regard are Taguchi's (1976) observations in which C: chla, P_m^{c} and α^{chl} vary widely.

Table 9.3. Interspecific variability of growth rate within higher taxonomic groups of marine and fresh water phytoplankton.

	Light : Dark Period	*T,° C*	*E*	*Max/ Min*	*c.v., %*	*n*
Cyanobacteria	C	20	60	5.1	38	22
Diatoms	C	18	320	3.5	49	6
Diatoms	C	21	8	1.6	19	5
	C	21	16	1.5	16	5
	C	21	32	1.2	8	5
	C	21	80	1.4	13	5
	C	21	160	1.4	15	5
	C	21	256	4.9	43	5
Diatoms	C	24	35	–	84	10
	C	24	80	–	79	9
	C	24	350	–	96	10
	C	24	800	–	135	10
	14:10	24	35	1.9	18	10
	14:10	24	80	1.9	19	9
	14:10	24	350	2.5	27	10
Dinoflagellates	C	21	8	4.2	54	5
	C	21	16	2.4	42	5
	C	21	32	3.8	46	5
	C	21	80	3.9	45	5
	C	21	160	–	79	5
	C	21	256	–	120	5
Dinoflagellates	C	24	35	–	87	6
	C	24	80	–	64	
	C	24	350	–	84	7
	C	24	800	–	125	7
	14:10	24	35	–	79	
	14:10	24	80	6.8	53	6
	14:10	24	350	–	65	7
	14:10	24	800	–	115	6
Coccolithophorids	C	24	35	–	67	4
	C	24	80	–	68	4
	C	24	350	–	70	4
	C	24	800	–	70	4
	14:10	24	35	2.3	32	4
	14:10	24	80	2.0	28	4
	14:10	24	350	2.8	38	4
	14:10	24	800	3.3	51	4

Green aglae	C	16	10	1.2	11	4
	C	16	25	2.0	26	4
	C	16	40	2.8	42	4
	C	16	60	2.4	39	4
Green algae	C	26	10	1.7	22	4
	C	26	25	2.5	43	4
	C	26	40	2.6	53	4
	C	26	60	4.8	54	4

Table 9.4. Interspecific variability in photosynthesis and chlorophyll *a* biomass rations within higher taxonomic groupings of marine and freshwater phytoplankton.

Taxonomic group	*T, °C*	*E. μmol. $m^{-2}. s^{-1}$*	*Max/ Min*	*c.v., %*	*n*
Cyanobacteria					
P_m^{chl}	20	30	4.7	27	20
P_m^{chl}	20	150	2.3	27	20
α^{chl}	20	30	1.7	18	20
α^{chl}	20	150	1.9	20	20
Diatoms					
P_{chl}	18	320	5.8	52	6
C:chla	18	320	1.7	22	6
C:N	18	320	1.4	13	6
Diatoms					
P_m^{chl}	5-8	50	52.0	142	16
α^{chl}	5-8	50	13.0	65	16
C:chl*a*	5-8	50	4.5	54	16
Diatoms					
chl*a* : protein	21	8	2.0	27	5
chl*a* : protein	21	16	1.6	18	5
chl*a* : protein	21	32	1.4	12	5
chl*a* : protein	21	80	1.3	18	5
chl*a* : protein	21	160	2.2	31	5
chl*a* : protein	21	256	3.2	43	4
Dinoflagellates					
chl*a* : protein	21	8	1.5	25	5
chl*a* : protein	21	16	2.4	45	5
chl*a* : protein	21	32	2.4	28	5
chl*a* : protein	21	80	2.1	29	4
chl*a* : protein	21	160	–	70	5
chl*a* : protein	21	256	–	104	5

Higher taxonomic affiliations can account for a large fraction of the interspecific variability noted above. It is commonly observed that dinoflagellates grow slower than diatoms when cultured under identical environmental conditions, whereas chlorophytes and coccolithophorids appear, on average, to grow at rates comparable to those achieved by the diatoms.

Organism size also appears to account for a significant fraction of interspecific variability. Generally the rates of metabolic processes in both unicellular and multicellular organisms have size dependencies that can be described by an allometric relation of the form

$$\text{rate} = aW^b \qquad ...(14)$$

where *rate* is the rate of some metabolic process, W is a measure of organism size, and *a* and *b* are constants. When specific metabolic rates with units of inverse time are plotted against mass, the exponent of Eq. (8) usually varies from –0.25 to –0.33. In other words, specific metabolic rates decrease with increasing organism size. Within the miroalgae, the allometric dependence only becomes evident when organisms from a wide range of cell size are considered. Groups of organisms that meet these criteria and for which sufficient data is available to examine the size dependence of metabolic rates, include the diatoms, dinoflagellates, chlorophytes and cyanobacteria.

Growth rate is one of the few variables for which a size dependence has been examined in microalgae. When data from all classes of microalgae are included in allometric equations, growth rate scales approximately with mass as W-0.30 apparently consistent with observations for other groups of organisms. However, when data are grouped by higher taxonomic affilation, the size dependence is markedly reduced. Schlesinger *et al.* (1981) provide evidence for a reduced size dependence of growth under suboptimal environmental conditions previously postulated by Banse (1976). Significantly, organism size can account for approximately 75% of the variance in growth rate.

Bulk biochemical composition shows little size dependence within particular classes of microalgae although it may vary significantly between higher taxonomic groups. In particular, dinoflagellates are characterized by lower chlorophyll *a* : biomass ratio than diatoms but it is unclear how relative pigment concentrations vary among the other algal classes or in the cyanobacteria. There is some evidence for a reduction in the ratio of chlorophyll *a* to biomass as cell size increases but this reduction in pigmentation does not appear to be sufficient to offset a reduction in chlorophyll *a* -specific photosynthesis rates (P*chl*) due to a reduced growth rate.

The size-dependence of growth rate discussed above should not be allowed to overshadow the potential use of other functional attributes as predictors of algal growth and photosynthesis, especially in the light of the weak size dependence often observed for specific growth rate. It is well established that a significant proportion of phenotypic variation in growth can be accounted for by differences in the cholorophyll *a*-to-carbon ratio and irradiance. However, the use of variables such as the chlorophyll *a*-to-carbon ratio to describe inter specific variations in μ are less well established. An important exception is the observation by Chan (1978) that most of the variation in nutrient-sufficient specific growth rates of diatoms and dinoflagellates could be accounted for by differences in irradiance and the chlorophyll *a*-to-protein ratio. These results indicate that algal growth rates can be predicted by considering the combined effects of one environmental variable and one physiological variable. Specifically, growth rate was found to be linearly related to the chlorophyll *a*-to-protein ratio at each of the six irradiances examined by Chan (1978). Underlying this relationship between μ, and the chlorophyll *a*-to-protein ratio is an implicit limitation on the interspecific variability in the parameters P_m^{chl} and α^{chl} of the photosynthesis-light response curve.

Problems in measuring Parameters of the Photosynthesis-Light Response Curve

Although often considered to be an attribute solely of the organism being examined under a given set of environmental conditions, the PI curve is not fixed exclusively by these factors, but also depends in a complex manner on the duration of the experiment. Two extremes in duration can be imagined. At one extreme, the response of growth rate to irradiance is obtained after exposure to continuous, usually constant irradiance, for a period of days to weeks. At the other extreme, the response of the water splitting reaction of PS2 to irradiance can be obtained through flash yield experiments in which individual flashes of < 1.μs duration are used to probe the operation of PS2 photochemistry will have different dependencies on irradiance due to differing rate limitations and regulatry activities although direct comparisons of the light-response curves for PS2 activity and growth have not been undertaken.

The Appropriateness of Current Protocols : a question of time Scale

Most determinations of the PI curve employ experimental durations of minutes to hours, a range of time scales that is clearly intermediate to the two extremes mentioned above. Transients in the O_2 evolution rate that accompany the onset of illumination, or a large change in irradiance, are well characterized and are typically complete with in a period of second. The steady rate of O_2 evolution over a period of minutes that follows these transients is often used indeterminations of the PI curve. Myers (1970) was the first to discuss the difference between the photosynthesis-light response curves obtained from these nearly instantaneous observations and the longer term dependence of growth rate on irradiance. When expressed in the appropriate units, response curves for growth and

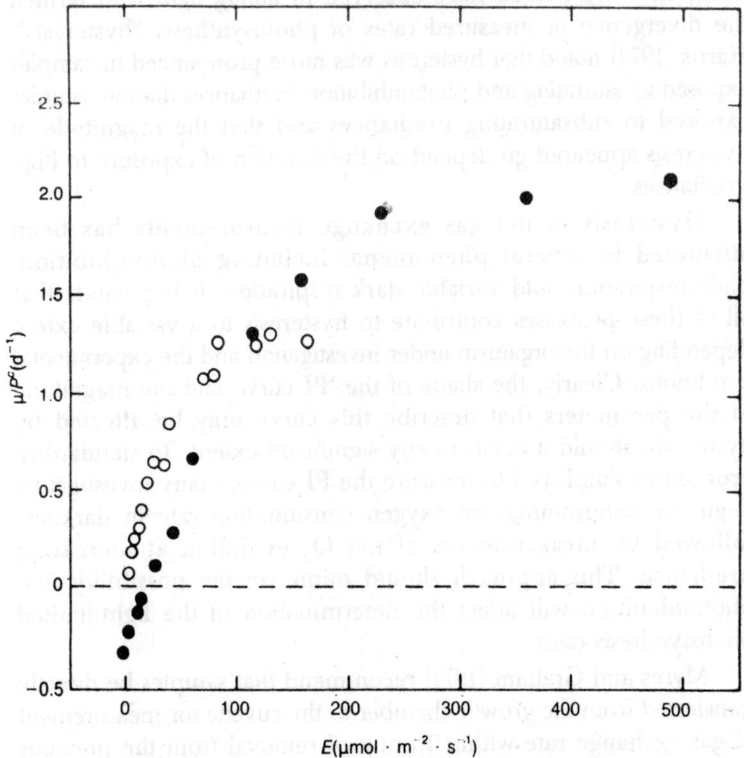

*Fig. 9.5. **Comparison of the photosynthesis-irradiance and growth-irradiance curves of the marine diatom Phaeodactylum tricornutum.***

photosynthesis are expected to cross at the irradiance to which the algae are acclimated.

Comparisons of PI and growth irradiance response curves are difficult to make because of differences in the geometries of culture vessels, light sources and methods employed to measure growth and photosynthesis. A recent comparison of *Phacodactylum tricormutum*, where it can be observed that the instantaneous photosynthesis rate exceeds the growth rate at irradiances above that to which the organism was acclimated, but that the growth rate exceeds the photosynthesis rate at lower irradiances. Even when PI curves are obtained in short-duration experiments, it is found that the rate of photosynthesis at a given irradiance may depend on the previous irradiance and Harris and Piccinin (1977) called attention to the problem that instantaneous rates of photosynthesis observed in rising light my exceed rates observed in falling light, and termed the divergence in measured rates of photosynthesis "hysteresis", Harris (1973) noted that hysteresis was more pronounced in samples exposed to saturating and photoinhibition irradiances than in samples exposed to subsaturating irradiances and that the magnitude of hysteresis appeared go depend on the duration of exposure to high irradiances.

Hysteresis in net gas exchange measurements has been attributed to several phenomena, including photoinhibition, photorespiration, and variable dark respiration. It is possible that all of these processes contribute to hysteresis to a variable extent depending on the organism under investigation and the experimental conditions. Clearly, the shape of the 'PI curve and the magnitude of the parameters that describe this curve may be affected by hysteresis, should it occur to any significant extent. To standardize procedures employed to measure the PI curve, many investigators begin by determining the oxygen consumption rate in darkness followed by measurements of net O_2 evolution at increasing irradiance. This approach should minimize the possibility that photoinhibition will affect the determination of the light-limited photosynthesis rates.

Myres and Graham (1971) recommend that samples be directly transferred from the growth chamber to the cuvette for measurement of gas exchange rate within 10 min of removal from the previous growth condition. This avoids any influence of previous treatments on measured photosynthesis rates, but the precision of the analysis may suffer since measurement of oxygen exchange are made on separate samples taken at separate times.

"Determination of the PI curve using ^{14}C-bicarbonate assimilation employs parallel incubation of a set of subsamples at a range of irradiances. Thus, each subsample is exposed to a different irradiance for the same time interval, which may range from minutes to hours. Sample size can vary from 1 cm^{-3} incubated directly in a scintillation vial to a more typical volume of 50 to 200 cm^{-3}. The preceding discussion indicates that the photosynthetic response of algae may not always be unambiguously described by PI curves. The utility of the PI curve decreases for algae exposed to very high irradiances for an extended period of time. In particular, photosynthesis rates decline during long exposure to photoinhibiting irradiances, with both the initial rate and the rate of decline dependent on irradiance. The rate of decline and any subsequent recovery depends not only on the experimental irradiance, but also on the the physiological condition of the algae under investigation. As a consequence, the parameters of the PI curve are seen to depend on the duration of the experiment as well as on the characteristics of the sample.

10

PHOTOSYNTHESIS IN NATURE

Of all aspects of plant metabolism, photosynthesis shows the most prominent variation under the dictates of the immediate environment. This sensitivity, and the dominance of the process in the synthetic economy of nature, has stimulated its study under more or less natural conditions during the past forty years. Nevertheless, a majority of physiologists have regarded with disfavour the unstandardised and arbitrary limitations of "natural conditions"; such experimental conditions are generally not easily attained, even in field experiments. There appears to be no comprehensive and coherent organization of the data now available. The principal difficulties lie in the incomplete integration of laboratory and field observations, work on terrestrial and aquatic vegetation, and measurements of photosynthesis and growth yields. Studies on land plants developed rapidly in 1918 and the years immediately following, particularly those by Boysen-Jensen Henrici Lundegardh, and Stalfelt. They benefited especially from the kinetic foundations laid by Blackman and his school in earlier decades and by Willstatter and Stoll in 1918.

The unique value of aquatic plants, particularly unicellular algae, was less readily appreciated for illustrating photosynthetic processes in nature than for the laboratory work on the photosynthetic mechanism that followed the pioneer experiments of Warburg. The use of these organisms developed from field experiments by marine and freshwater biologists in the 1920's and has recently shown a spectacular expansion in connection with measurements of the photosynthetic productivity of plankton. The peculiar problems presented by photosynthesis in nature chiefly concern the kinetics of the process.

The photosynthetic rates in any vegetation cover are controlled by a considerable number of internal and external factors and by interactions between these factors and apparently order little prospect for a unified treatment. However, light, as the energy source, can be regarded as the master factor, and shows more remarkable variation than any other environmental factor. Much behaviour can be interpreted in relation to the rate-intensity characteristic of photosynthesis, its modification in response to other factors, and its expression in the variation in space and time of photosynthetic rates under natural conditions. This characteristic also expresses the two universal limitations of the efficiency of energy utilization : the maximum quantum yield at low intensities, and light-saturation at high intensities. It consequently provides a starting-point for comparing observed with potential yields.

The conversion of net photosynthesis to gross photosynthesis by correcting for respiration losses is often uncertain, partly because of possible changes in respiration during illumination therefore, many field measurements of photosynthesis are expressed as net rates only. Where possible, however, this discussion will be based upon the probable gross photosynthesis, with the consequences of respiration introduced separately.

Sources of Information

In most field work small samples of plant material have been used, exposed in some form of transparent chamber, cuvette, or bottle. Here the photosynthetic uptake of carbon dioxide or liberation of oxygen could be measured either directly or following the absorption of residual carbon dioxide from an air-stream in alkali solution. Changes due to carbon dioxide uptake by both terrestrial and aquatic plants have been estimated by differences in titration, electrical conductivity, pH, and–in recent years–by fixation of radiocarbon(^{14}C). Infrared absorption analysis, which enables the sensitive and continuous recording of carbon dioxide in air, has been much exploited recently in work on land plants. The evolution of oxygen is readily measured only with aquatic plants, for which the Winkler method has been widely used.

A brief survey of methods employed for terrestrial plants in given by Frenzel and for aquatic plants by Lund and Talling. Work with entire terrestrial plants or stands usually requires larger and more elaborate plant chambers; examples are described by Boysen-Jensen Heinicke and Childers and Thomas and Hill. Any enclosure involves a danger of the development of special conditions of the *Kuvettenklima*, particularly by overheating in air and reduced mixing in water.

Various tests of these effects, and some suggested remedies such as improved circulation of cooling jackets, have been described. These difficulties are absent, but others–especially those attributable to translocation–are present in work based on the measurement of assimilates formed by leaves. Field results generally fall in two main groups that are governed by the differing opportunities for exposures between terrestrial and aquatic vegetation. With the former, the diurnal variation of photosynthetic rate has been studied intensively by successions of short exposures, but the vertical variation of rates in the leaf canopy is rarely investigated.

For aquatic plants the variation of rate with depth has been the centre of study, with diurnal changes often obscured in long exposures. Besides these complementary deficiencies, the data often permit only qualitative, or semi-quantitative correlations to be made between the variation of photosynthetic rate and the more important environmental variables such as light intensity, temperature, and carbon dioxide supply. Advantages offered by a recourse to laboratory exposures are here obvious. In the first instance, such experiments allow the comparison of some photosynthetic indices using plant samples collected in different situation. Examples are provided by studies of the variation of photosynthetic capacity with latitude or season and of differentiation between "sun" and "shade" material.

The application of such measurements to a reconstruction of photosynthetic activity in nature requires more extensive sampling and measurement of environmental variables than is usually made; direct comparisons of laboratory and field exposures (*e.g.,*) are rarely attempted. This "synthetic" approach to the estimation of photosynthetic activity under natural conditions was most strikingly developed by Boysen-Jensen and his associates. Their contributions are among the most stimulating to the subject, although some oversimplification was inevitable and was criticized by others. Some attempts have been made to avoid the limitations of small enclosed samples by estimating the total photosynthesis of a plant community from the diurnal changes produced in the gaseous composition of the ambient medium. For small water bodies the changes are often confined in horizontal and vertical directions and are relatively accessible for study. Measurements for a representative water column of dissolved oxygen and carbon dioxide have been used; the chief uncertainties arise from the variable exchange between water and atmosphere.

The diurnal changes of atmospheric composition produced by land vegetation are more susceptible to turbulent mixing and dissipation, but in favourable weather conditions can follow a regular course and affect an air-column of considerable height. Their use in estimating photosynthetic fixation of carbon dioxide is discussed by Huber. Lastly, the carbon increase indicated by plant growth can be regarded as an integration of net photosynthesis with time. However, there are few quantitative comparisons with gasometric data, although such are invited by one particularly successful technique of growth analysis.

Potential and Actual Yields

The simplest model of a photosynthetic cover of vegetation would be provided by an uniform algal suspension, absorbing all incident light and operating at the maximum quantum yield of photosynthesis. In illustration one may assume an average of 0.1 molecules oxygen per quantum for the latter corresponding to an average efficiency of solar energy utilization in the spectral region of 400 to 700 mμ, of about 22 per cent. The photosynthetically available energy in this spectral region normally accounts for a fraction, close to 0.46, of the total energy received. Consequently, a radiation total of 600 cal./ cm^2. (horizontal surface), common on very sunny days at many latitudes, would lead to a photosynthetic production of carbohydrate that was equivalent to 65g.C/m^2 per day. This value is a measure of the gross activity of an ideally efficient vegetation cover under conditions of radiation a little below the maximum experienced on the earth. With it, the yields of less efficient models of vegetation, as well as yields actually observed, can be usefully compared.

Two other model situations, readily visualised, may be founded on the experience of photosynthetic rates measured per unit leaf area. At lower light intensities, in which the rates increase linearly with increasing intensity, increases of 1.5 to 2.0 mg. CO_2/100 cm^2. per hr. in 1 kilolux increments of intensity are often observed. If a daily energy total of 600 cal./cm^2 (approximately 770 kilolux-hours) were entirely utilised in this way, in a completely absorbing leaf cover, a fixation of 31 to 42 g. C/m^2 per day would result. This figure is again an upper limit of yield, as light-saturation of photosynthesis is likely to be reached during a considerable part of the supposed sunny day. One method for estimating its effects is proposed by de Wit. Photosynthetic rates of healthy leaves at light-saturation commonly lie between 10 and 25 mg. CO_2/100 cm^2. per hr, in air not enriched with CO_2. If these persist during a 12-hour day with a uniform and uninter rupted vegetation cover that is one leaf in thickness, the corresponding fixation will be 3.3 to 8.2 g C/m^2. per day.

Values of the actual yields of vegetation can be based upon measurements for short periods (*e.g.,* one or several days) or much longer times (*e.g.,* a growing season, or a year). It was noticed long ago in Europe *e.g.,* Mayer that the annual dry weight yields of many dense tree and crop stands were of remarkably similar magnitude, usually 5000 to 10,000 kg./hectare. Later experience has generally confirmed this magnitude, but has also indicated some appreciably higher yields (15,000 to 35,000 kg./hectare). If the total dry weight contains an average carbon content of 40 per cent (values commonly lie between 30 and 50 per cent), the higher range quoted above corresponds to average net fixation rated of 1.6 to 3.8 g. C/m^2 per day in a maximum growing season of 365 days. Owing to respiration and other losses, the corresponding gross photosynthesis is often likely to be 50 to 100 per cent higher *e.g.,* Heinicke and Childers Moller *et al.* The gross photosynthetic yields of dense natural populations of planktonic algae, measured by gas exchange during short periods, reach values of 3 to 4 g. C/m^2 per day.

These yields are clearly far below those permitted by the maximum quantum yield of photosynthesis. The gap is equally evident in numerous other comparisons bases upon percentage energy conversion over long periods. However, recent work with both algae and terrestrial plants has shown that, in short periods under favourable field conditions, the gap between actual and maximum potential yields can be considerably reduced. Improved growth vields in algal mass cultures under natural illumination can represent an overall energy conversion (spectral region 400 to 700 mμ) of around 6 per cent and a net carbon fixation of approximately 5 to 10 g./m^2 per day; and the maximum daily yields of some crops.

The preceding estimates of yields in model situations suggest that much of the difference between actual and potential values originates from light-saturation of photosynthesis under natural conditions. It is also indicated by several analysis of actual yields. More specific expressions of this effect are discussed below.

The Expression of Light-Saturation under Natural Conditions

Rate-intensity curves of photosynthesis almost universally show a transition between a linear relation of rate and light intensity when the latter is low, and a maximum rate which usually persists over a considerable range of higher intensities. This transition to light-saturation is often gradual, especially in optically dense material but can be most precisely indicated by the intensity (I_k) at which a continuation of the initial gradient (θ) would reach the light-saturated rate (P_{max}). The more common visual estimates of the intensity at which light saturation occurs are generally about twice this, I_k value. The Ik index also has the advantage of linking P_{max} and θ characteristics by the simple relation $P_{max} = \theta.\ I_k$, and can be reinterpreted in terms of other mathematical formulations of the rate-intensity characteristic. A division of I_k by the ratio between P_{max} and respiration rate (R) (*i.e.,* P_{max}/R, related to the "economic coefficient" and "balance quotient" of many European authors). yields the value of the compensation intensity if this falls within the initial linear region of the rate intensity chracteristic.

The variation of I_k can be deduced from its two determinants, P_{max} and θ. The former is the most variable, and so accounts for most of the changes in I_k. Illustrative examples are provided by the sensitivity of P_{max} and I_k to temperature and stomatal aperture the depression of P_{max} being associated with reduced I_k values. It is therefore dangerous to interpret in isolation the variable onset of light-saturation, as is often done in the description of supposed heliophilic or ombrophilic characteristics. Thus Boysen-Jensen and Boysen-Jensen and Muller pointed out the frequent similarity of the initial gradient (θ) in sun and shade examples of different species and in leaves of the some species : here saturation at low light intensities in the shade forms was largely imposed by correspondingly low rates at light-saturation, assessed per unit area of leaf or thallus.

Measurements of photosynthetic rates in the field express the consequences of light-saturation in several ways. It is usually evident in diurnal curves of photosynthesis based upon small plant samples and in rate-intensity curves calculated therefrom. Another illustration is given by comparative measurements of photosynthesis below artificial or natural light screens of different densities. A pronounced light-limitation of photosynthesis is often shown only with day-light factors (the fractional or percentage illumination relative to full day-light) of < 30 per cent; such factors are often called "light intensities," but represent much more complex light conditions in varying natural illumination. In the most interesting situation of this kind, best illustrated by work with planktonic algae, the light screen is provided by varying depths in the community under study. Light-saturation can then be shown in a depth zone with maximum photosynthetic rates, passing below into a near-exponential decline of rates with increasing depth, and often ending above in a superficial zone with depressed rates associated with strong illumination.

Such depth-profiles can be related to basic photosynthetic characteristics by analysis, and to the overall rate under unit area of cover by integration. Thus they afford a means of connecting the integrated yields per unit are with basic photosynthetic characteristics. The maximum rates possible in these profiles, some special problems linked with their upper and lower regions, the thickness or density of the plant cover through which they extend, their integration in relation to depth and time, and comparison with an alternative approach of growth analysis are discussed below.

Variation of Photosynthetic Rates at Light-Saturation

Photosynthetic rates measured at (or near) light-saturation can be used to compare the absolute capacities of plant material differing in type and origin, and to illustrate the effects of factors with ecological significance.

Ecological consequences can be anticipated for the well known increase of light-saturated rates with temperature, depression and inhibition at very high and low temperatures, sensitivity to previous thermal history and the effects of the temperature-dependence of respiration rate upon net photosynthesis. The abundant records of daily photosynthesis by land plants seldom indicate any strong positive influence of diurnal temperature change. Demonstration of such influence is also hindered by the problems of measuring true leaf temperatures and allowing for over-heating effects in small chambers. These difficulties are much reduced with water plants, for which Gessner describes some probable effects of diurnal temperature change. Diurnal changes of temperature are usually much greater in terrestrial than aquatic environments. In the former, high leaf temperatures, especially in the afternoon, are often accompanied by depressed rates of photosynthesis. Besides the direct depression of photosynthetic rates at high temperatures several indirect consequences of temperature, affecting water deficit and respiration rate, can be involved.

Some pronounced examples of the diurnal or seasonal depression of net photosynthesis, which could be related to the rise of respiration rate in warmer conditions, are recorded for desert plants for various mosses and lichens and for conifers. In the algae, laboratory studies of some thermophilic strains of *Anacystis* and *Chlorella* show the utilization of high temperatures for achieving remarkably large rates of photosynthesis and growth and noticeably high rates may also be found in tropical freshwater plankton under field conditions. The many rates recorded for phytoplankton from other regions, assessed per unit chlorophyll content, are usually in the range of 3 to 15 mg. O_2/mg, hr. (or its equivalent) at light-saturation Tropical examples of terrestrial plants studied do not appear to show particularly high photosynthetic rates at light saturation.

The effects of low temperatures upon the photosynthetic activity of land plants in nature have been much investigated, particularly in relation to the performance of evergreen plants in winter. There is general agreement that significant, if reduced, photosynthesis can occur at air temperatures below 0°C. Tranquillini describes development of the photosynthetic capacity of leaves of *Pinus cembra* still covered in alpine snow.

The limitation of photosynthetic rates in nature by the carbon dioxide supply is generally agreed to be most pronounced in terrestrial habitats, where the atmospheric concentration of carbon dioxide is low and averages about 0.03 per cent or 0.59 mg./1. near sea level. Deviations from the mean have been studied extensively. The physiological significance of the most regular form of variation, the decrease with altitude, is little understood. Despite the low absolute–though not relative–carbon dioxide concentrations at high altitudes, there are recurrent reports of unusually high photosynthetic rates by alpine plants in nature. Daily changes of concentration are also very common in and near vegetation and often are measured in records of the diurnal course of photosynthesis. However, a pronounced correlation between concentrations and rates is rarely evident and the total daily variation of concentration is usually not large, though an air column of considerable thickness may be involved. High concentrations under a canopy near the soil have been supposed to offset the less favourable light conditions there, but direct proof is difficult and in some cases the evolution and accumulation of carbon dioxide was probably overestimated. Wind movements have an obvious importance for such distribution patterns, and there are some quantitative data for the effects of low velocities on the supply of carbon dioxide to photosynthesizing surfaces. The stomatal control of carbon dioxide supply has been much discussed in relation to midday or afternoon depression of photosynthesis. Variation in stomatal aperture probably also accounts for some differences in the photosynthetic activity of leaves at different levels in vegetation and possibly for some seasonal differences.

There are fewer data for acquatic plants on the relation between carbon dioxide supply and photosynthetic rates in nature. The limitations of diffusion to massive plant organs in water are well known, so that water currents and turbulence are likely to be of considerable ecological significance. Such indications are generally lacking for the dilute natural suspensions of planktonic algae with a much greater surface to volume ratio. The photosynthetic removal of carbon dioxide from the natural carbonate-bicarbonate buffer system is reflected in a rise in pH. Many diurnal examples have been described *e.g.*, and some were used for the estimation of daily photosynthetic productivity. However, a severe limitation of photosynthesis by carbon dioxide depletion is only likely at pH values higher than are commonly found. A probable example for freshwater plankton algae is described by Steemann Nielsen. Some species, such as *Arthrospira platensis* show a remarkably dense development and active photosynthesis in very alkaline waters with pH exceeding 10.5. There is a considerable amount of evidence for the utilization of bicarbonate as carbon dioxide source by many aquatic plants.

Seasonal changes of photosynthetic capacity, which reflect the previous history of the plant material, have been described from both field exposures and from standardized laboratory measurements combined with periodic sampling of vegetation. The first frosts of winter lead to a sharp fall in the maximum photosynthetic rates of many land plants. In evergreens relatively low photosynthetic capacities are usually found in the winter period although recovery may occur in a few days in warmer conditions. The variation of photosynthetic activity with age of individual leaves is also a source of seasonal variation. Seasonal rains in hot dry regions can produce a remarkable rise in photosynthetic rates although- as in tidally exposed seaweeds the dry phase need not represent a "resting season" for photosynthesis (73).

A diurnal depression of photosynthetic capacity has been very commonly found for land plants, and is also recorded for aquatic macrophytes. In recent years examples have been described for marine and freshwater phytoplankton, with maximum capacities usually reported near dawn and minimum near sunset. This daily decline may be connected with the photo-inhibition often-found in short exposures near the water surface; Shimada gives evidence for parallel changes in chlorophyll-*a* content. Behaviour in terrestrial plants is more varied, often with slight to large depressions near midday, persisting in some examples throughout the afternoon. The occurrence of these patterns in the same species at different alotades is described by Zatensky. No single explanation can fit all the examples described ; there is evidence for the effects of stomatal control [e.g., Nutman] the direct inhibition of photosynthesis at high temperatures and light intensities [cf.Stalfelt], accumulation of assimilates[e.g., Guttenberg and Buhr Kurzanov water deficit and–for net assimilation–increased respiration rates at higher temperatures. As almost all the data are derived from diurnal exposures in natural illumination, it is difficult to separate the effects of depressed rates at light-saturation (attributable to the previous history of the material) from the unfavourable influence of the immediate environment. Kostychev and his co-workers stressed the importance of internal factors in determining large diurnal fluctuations of photosynthetic rate, but these views were not supported by much other work.

Several comparisons have been made of the maximum photosynthetic rates shown by plants from various terrestrial habitats or climatic zones. The first field survey of this type on a large scale was undertaken by Kostychev and his associates. The high rates described for various alpine plants and the comparatively modest rates for some tropical plants have already been mentioned.

It appears that high rates of photosynthesis are often achieved in apparently unfavourable habitats with a short growing season the brevity of which is thereby partly compensated; there are examples of this for alpine, arctic and desert plants. Continuous photosynthesis under arctic summer conditions is describe by Kostychev *et al.* and Kisliakova considered that the variation of photosynthetic capacities of species in continental climates, as in Central Asia, was greater than in the floras of more oceanic or tropical climates. Succulent halophytes do not appear to show marked peculiarities of gross photosynthetic behaviour, although rates expressed on a fresh weight basis are naturally low and the ratio of rates of respiration to photosynthesis on a unit area basis may be unusually high. Work on one halophyte, *Aster tripolium* provides one of the few clear examples of different photosynthetic responses between ecologically distinct forms of the same species.

Some Limitations of Photosynthesis in Exposed and Shaded Situations

These limitations are shown most diagrammatically in many of the depth profiles of photosynthetic activity reported for planktonic algae, and have been noted briefly. The depression of photosynthetic rates near the water surface is common at light intensities exceeding 100 kiloerg (400 to 700 mμ)/cm^2 per sec. and a depth corresponding to 1 percent of the surface intensity is usually taken as the lower limit of the zone in which appreciable photosynthesis can occur. The overall significance of the surface depression is very uncertain in relation to algal populations circulating over a greater depth range; it may affect subsequent rates of photosynthesis at light-saturation. There is also some evidence for a damaging influence of radiation, such as the ultra-violet, rapidly absorbed by the water. The reduced photosynthesis often shown near the water surface, when the algae were sampled from the depths at which they were later exposed may often be caused by decreased population density rather than specific rates of photosynthesis.

The decline of photosynthetic rate at greater depths reflects the exponential decline of light intensity. It bears no constant relation to the relative or percentage light values, but is also determined by the logarithmic interval between the surface light intensity (I_o) and the intensity that measures the onset of light-saturation (I_k). The greater this interval, the deeper the depth-profiles are likely to extend. This source of variation is dwarfed, however, by the enormous range of light penetration in natural waters which leads to a variation of photosynthetic zones from 100m, or more in the clearer oceans to 1m or less in highly coloured and turbid waters.

The comparative photosynthetic rates of aquatic macrophytes at different depths have been studied less, but similar restrictions at shallow and great depths are known in fresh waters and in the sea. Attached algae and mosses have been found growing at depths which probably correspond to about 1 percent or less of the effective surface radiation although few adequate measurements of light penetration are available. The significance here of adaptation to the colour or intensity of light has been much discussed.

The light field is usually more heterogeneous in a terrestrial habitat than under water, and the consequences of a vertical gradient for photosynthesis are less easily demonstrated. Measurements of such gradients are discussed by different scientists but most light data apply to a single stratum in the plant cover, such as a forest floor. As under water, the 1 percent light level often marks the lower limit of appreciable photosynthetic activity, and values of this order are common at the soil surface beneath dense vegetation. The light level is also likely to represent the region of the compensation point during short exposures, as intensities of 20 to 80 kilolux are typical of average daily illumination with 0.2 to 0.8 kilolux for the compensation intensity.

Some extreme shade plants can grow at lower relative intensities, some measured as below 0.1 percent. Biebl describes a particularly interesting example for the moss *Mnium serratum* inhabiting rock-cavities, where the deepest shade tolerated was given as 0.024 percent, with a midday maximum of light intensity at about 26 lux. On exposure to full sunlight, death occurred within a few hours. Under a vegetation cover the assessment of the effective light climate is rendered more difficult by spectral modifications due to the leaf canopy the short-period fluctuations, sun-flecks and long-period changes connected with leaf development and fall. Botanists studied in detail the depression of photosynthetic activity that followed development of the tree canopy.

Some species, such as *Anemone nemorosa* showed a change from a positive to a persistent negative balance of net photosynthesis, and in others e.g., the positive balance was regained after a shift in the compensation point. Various factors that can reduce the light-saturated photosynthetic capacity are often most pronounced at the upper exposed level of a vegetation cover. A detailed exploration of the distribution of photosynthetic activity in the tree-crowns of *Picea excelsa* and *Fagus sylvatica* was made by Pisek and Tranquillini. Here reduced photosynthesis could be found associated with water deficit and stomatal closure in exposed shoots subjected to greater insolation. Photosynthetic activity extended most deeply on the more illuminated side, particularly at low solar elevations. Earlier, Filzer showed differences of photosynthesis between north and south aspects of trees that were connected particularly with light intensity, and others between east and west aspects with diurnal timing.

The significance to photosynthesis of the angular distribution of light has been little studied in either aquatic or terrestrial habitats, although the ecological application of photometeres recording spherical illumination have been described for both habitat types. Stocker has emphasized the probable importance for photosynthesis of the diffused reflection by leaves in some tropical vegetation. The differentiation of photosynthetic and other characteristics between plant individuals, or their parts, in exposed and shaded situations is very common, but there is less agreement on its adaptive significance. "Shade plants" and "shade leaves" are discussed by Boysen-Jensen and Lundegardh. They often show relatively low rates of respiration and maximum photosynthesis, with light-saturation and the compensation point found at low intensities. The last two features have elicited the most comment, but can usually be interpreted in terms of the first two in conjunction with a less variable initial gradient of the photosynthesis-intensity curve. Examples of woodland plants have been particularly studied in this connection.

The differentiation between sun and shade leaves on the same plant is most fully described for the beech. *Fagus sylvatica.* Differentiation between species was well illustrated in a survey of eight "sun" and five "shade" plants by Bohning and Burnside although the authors emphasized that the ecological significance is not simple, except in habitats with very low light intensities. The differentiation of "sun" and "shade" characteristics of photosynthesis has been described recently in relation to depth in phytoplankton communities the samples of deeper origin showed light saturation at lower intensities and a greater susceptibility to photo-inhibitation at higher intensities.

Ontogenetic adaptation or lability in relation to light intensity, leading to a differentiation of "sun" and "shade" characteristics, is known for both algae and for higher terrestrial plants. These modifications were studied by Wassink *et.al.* in leaves of *Acer pseudoplatanus* that were exposed to various intensities of artificial illumination and fractions of full natural illumination. Changes in the two series were qualitatively similar, with an increase in the light-saturated rate of photosynthesis and the onset of light-saturation with rising light intensity during adaptation , and a corresponding decrease in the chlorophyll content per unit area. Ontogenetic differentiation of photosynthetic characteristics has also been described by Burnside and Bohning and by Bordeau and Laverick for several species of conifer seedlings. The latter emphasize that exceptions to the typical differentiation of "sun" and "shade" behaviour occur. Shade leaves usually show a high chlorophyll content per unit weight, though–owing to their frequent thinness–not always per unit area.

The Density of Photosynthetic Cover in Nature

The density of photosynthetic material in the unit are of a natural habitat is an essential link between data on the photosynthetic behaviour of plant samples and the overall photosynthesis of the plant cover. However, systematic and comparative measurements in relation to photosynthesis under natural conditions are comparatively few.

In some respects the simplest situation is found in natural populations of planktonic algae, whose chlorophyll content (particularly chlorophyll-*a*) has been much studied in recent years. Most measurements relate to the content per unit volume of water; and although any limits are arbitrary, values of <1 mg/m^3 are common in the unproductive open oceans 1 to 30 mg/m^3 in moderately productive waters and > 30 mg/m^3 in very productive regions. A few calculations have been made of the content below unit area in a complete water column. Gessner estimated that in productive waters the chlorophyll content could reach values of about 1 g/m^2, similar to those found in closed terrestrial communities. The more relevant content, however, is that of the illuminated part of the water column (euphotic zone) in which appreciable photosynthesis is possible. The maximum content will depend on the extinction associated with unit density of the algal suspension. Steemann Nielsen used absorption data of chlorophyll extracts in organic solvents to estimated roughly the maximum content as about 300 mg/m^2, which appeared compatible with actual amounts per unit area in dense natural and cultured populations. It is also supported by observations on the increase of the light extinction during the growth of planktonic diatom in nature. The increment of optical density (In I_o/I) associated with unit density of algae then appeared low enough (of the order of 0.02 per mg. chlorophyll-*a*/m^2) to render unlikely any strong optical incursion by the densities of phytoplankton most commonly found in nature.

In most terrestrial vegetation the situation is obviously very different, both as regards the optical importance and the density per unit area of photosynthetic material, and its usual display in flat laminae. Steemann Nielsen considers that the density of chlorophyll often reached per unit of ground area is significantly higher than the maximum value possible for phytoplankton in the euphotic zone. He connects this difference with the optical advantages offered by leaf arrangement and structure, particularly the combination of diffusing epidermis and light channels in palisade cells. The predominant laminar arrangement allows the density of photosynthetic cover to be conveniently expressed by a pure number, the ratio of leaf area to ground area. In recent years this measure–the leaf area index–has been widely used in analyses of crop growth light interception by vegetation comparisons of plant cover and in theoretical treatments of photosynthetic productivity. It was also involved in earlier Scandinavian work on terrestrial "Stoff produkton". Watson gives an excellent general review in relation to crop plants, and some ecological aspects are treated by Walter.

In continuous plant cover almost all values of leaf area index fall between one and eight; low values of about one are probably common in shade habitats and higher values of up to eight are illustrated by some European beech woods. The upper limit of the index is clearly connected with the effective light extinction associated with each leaf layer. The relationship between leaf area index and vertical light extinction, indicated by Brougham, Monsi and Saeki, Nichiporovich and Chmora appears to be less strongly influenced by the average transmission (usually near 10 per cent) of individual leaves than by the type of leaf arrangement. With the very thin leaves of mosses the leaf area index can be correspondingly high. Light extinction is likely to be more directly related to leaf area index than to chlorophyll content per unit are, since much variation in the chlorophyll content per unit area of leaves (especially >5 mg/dm^2) has a small influence on either optical density or photosynthetic efficiency in weak light.

Depth-Integration of Photosynthetic Rates

The most complete assessment of the total photosynthetic activity below unit area of a plant cover would be provided by the summation of individual rates recorded at representative points within the cover. This approach has been applied extensively using the relatively homogenous communities of planktonic algae. Here the area enclosed by the depth profile photosynthetic rate. It is susceptible to the depth-limitations seen in exposures of uniform samples and also to the effects of any stratification that affects population density, composition, and differentiation of photosynthetic characteristics, between cells at various levels. The unravelling of these simultaneous variables has rarely been attempted. The effects of population density can be seen in the replotting of depth profiles with specific photosynthetic rates per unit quantity of pigments or in comparisons with crop profiles. The separation of depth changes as due to the vertical light gradient, from the differing photosynthetic capacity of population samples from various depths, has been used in several indirect methods of estimating photosynthesis per unit area.

Photosynthetic capacity has been measured for samples exposed in illuminated constant temperature baths or near the water surface with natural illumination. Little is known of the quantitative influence of a depth-differentiation of "sun" and "shade" characteristics upon the depth-integral of photosynthesis. The influence and interaction of individual variables also can be tested by a mathematical integration of photosynthetics rates, aimed at providing a theory that can be applied to analyse and compare actual examples. Formulations have centered upon the expression of the photosynthetic rate (P)–light intensity (I) characteristic, with maximum rate(P_{max}), in relation to the varying surface intensity (I_o) and vertical light intensity gradient, the latter defined by some "average" or "effective" value of the vertical extinction coefficient (k_e). As a limiting case a uniform population in the photosynthetic zone can be supposed, of density *n*. If the surface intensity I_o is sufficiently low so that light saturation of photosynthesis can be neglected, the formulation of the photosynthetic productivity per unit area (ΣP) is much simplified ; examples are given by Riley Syerdrup and Vollenweider. Berge has also used this simplification in computations for some sea areas at high latitudes. the more usual significance of light-saturation for ΣP has been calculated graphically by Ryther and in mathematical derivation by Talling. The central expression of Talling using the present symbols, can be reduced to

$$\Sigma P = \frac{nP_{max}}{k_e}\left(\ln \frac{I_o}{0.5I_k}\right)$$

This expression was used to formulate other factors involved in the photosynthetic utilization of solar radiation by planktonic algae, and was applied to the characteristics of some specific populations in nature. A parallel theory developed for the photosynthetic or growth kinetics of algal mass cultures has yielded expressions that can be re-interpreted in a similar form and applied over a wider range of the ratio I_o/I_k, namely

$$\Sigma P = \frac{nP_{max}}{k_e}\ln\left(\frac{I_o}{I_k}+1\right)$$

These equations give a formal and simplified expression to several limitations of the integral photosynthesis that have often been discussed. The relation to the logarithm of the surface light intensity implies that the integral photosynthesis will be relatively insensitive to variations in I_o when this (and the I_o/I_k ration) is high. Such insensitivity has been commented upon by various authors who have considered the vertical displacement and enlargement of photosynthesis-depth profiles with increasing surface intensity.

Derivation of the photosynthetic productivity in nature from measurements on samples is clearly more difficult for a terrestrial plant cover. Some comparisons have been made between measurements of the photosynthetic behaviour of entire stands and their component leaves rate-intensity curves and diurnal changes of photosynthetic rate show less abrupt saturation at higher it tensities in the dense stands. Transitional behaviour during stand growth is described for rice plots. Similar consequence of self-shading effects in dense leaf system have been discussed for various conifers. Several alternatives have been used in attempts to calculate indirectly the daily photosynthesis of a community in nature. As with phytoplankton they have centered on the combination of the rate-intensity characteristic with the daily variation of incident illumination. In examples given by Boysen-Jensen Larsen and Polster the characteristics from single leaves or twigs were used, and are not suitable for dense systems of overlapping leaves. Such a systems, utilizing the matted growth of the moss *Homalothecium sericium*, was used directly by Romose in his experiments. The results were combined with data on diurnal illumination at various seasons to estimate photosynthetic activity under natural conditions, using an average gradient for rateintensity curves in relation to daily illumination-integrals (as lux-hours).

A mean gradient was used in another way by de Wit to calculate theoretically the potential daily photosynthesis of dense plant cover from daily radiation-integrals; in the latter he-distinguished two fractions, one being utilized in the calculation. As in dense algal communities, the corresponding vertical light gradient can be related to the density of photosynthetic material. The stratification of the latter, and the associated high gradients, are illustrated for various natural communities, and discussed in relation to the transmission of individual leaves. The depth integration of rates yields an expression that can be transformed by appropriate subsitution of symbols into a form very similar to that quoted for algal suspensions. The significance of the ratio P_{max} to k_e (the latter now related to leaf area index), and of the logarithmic function of the ratio I_o/I_k, can then be seen.

The model is illustrated by data of Boysen-Jensen on the photosynthesis of leaves and plant stands. Further and more direct comparisons with observed photosynthetic rates are needed to test the assumptions and simplifications involved. Integrated yields of photosynthesis ~an be used for independent comparison with growth increments if both the carbon content of the latter and the various sources of loss (e.g., respiration) are kr.›wn quantitatively. Few studies of photosynthesis under field conditions are sufficiently detailed to permit the comparison, but included among the few are those of Heinicke and Hoffman and Heinicke and Childers on *Pyrus malus.*

Larsen on *Solanum* on *Homalothecium sericium* Thomas and Hill on several crop plants, and Tranquillini on *Pinus cembra.* In these exampies, except that of *Pinus cembra*, the agreement was good considering the problems involved, particularly the one of obtaining estimates of respiration losses over long periods. Uncertainities concerning these and other losses have generally prevented the effective comparison of growth and photosynthetic rates for natural populations of planktonic or other algae. The calculation of growth increments per unit of leaf quantity (usually area) is the basis of particularly successful form of growth analysis suitable for ecological application.

The instantaneous rate of growth, defined as the "net assimilation rate" or "unit leaf rate" (E), can be related to the unit area productivity of the stand by the more variable factor of the leaf area index, and to the relative (specific) growth rate by the ratio of leaf area to total dry weight. Average values of E during a growing season often show a remarkably limited variation with species or latitude, although this is now known to be considerably broader than the "typical" range of 0.4 to 0.7 g/m^2 per week originally suggested by Heath and Gregory Seasonal variation is described by Watson maxima of E may not coincide with the best development of leaf area index, thus limiting the unit area productivity.

Another limitation of unit area yield, by an inverse relation between E and leaf area index, can also develop the optimum leaf area index for yield is therefore of much interest. The average magnitude of E corresponds roughly with that expected from the usual order of the net photosynthetic capacity of leaves at light saturation, about 0.7 to 1.7 g carbohydrate/m^2 per hr. Although the net assimilation rate reflects factors other than the net photosynthetic rates of leaves (e.g., respiration of non-photosynthetic organs, ash content), the relation of these quantities is sufficiently close to make their combined study in field conditions most desirable. There is, for example, a suggestive parallel of a modified logarithmic relationship established between net assimilation rate and the daylight factor and a similar relationship is indicated by theory between photosynthetic productivity per unit area and light intensity in a dense homogeneous cover.

INDEX